THE DANCE OF
AIR AND SEA

THE DANCE OF AIR AND SEA

How Oceans, Weather, and Life
Link Together

ARNOLD H. TAYLOR

OXFORD
UNIVERSITY PRESS

OXFORD

UNIVERSITY PRESS

Great Clarendon Street, Oxford OX2 6DP

Oxford University Press is a department of the University of Oxford.
It furthers the University's objective of excellence in research, scholarship,
and education by publishing worldwide in

Oxford New York

Auckland Cape Town Dar es Salaam Hong Kong Karachi
Kuala Lumpur Madrid Melbourne Mexico City Nairobi
New Delhi Shanghai Taipei Toronto

With offices in

Argentina Austria Brazil Chile Czech Republic France Greece
Guatemala Hungary Italy Japan Poland Portugal Singapore
South Korea Switzerland Thailand Turkey Ukraine Vietnam

Oxford is a registered trade mark of Oxford University Press
in the UK and in certain other countries

Published in the United States
by Oxford University Press Inc., New York

© Arnold H. Taylor 2011

British Library Cataloguing in Publication Data

Data available

Library of Congress Cataloging in Publication Data

Data available

Typeset by SPI Publisher Services, Pondicherry, India
Printed in Great Britain
on acid-free paper by
Clays Ltd, St Ives plc

ISBN 978–0–19–956559–7

1 3 5 7 9 10 8 6 4 2

To my mother and father.

CONTENTS

LIST OF ILLUSTRATIONS

PREFACE

The journey to this book began in August 1990, when a student, Nick Baker, who was assisting me for a few months, constructed the pair of graphs which are reproduced in Fig. 1.1 (at that time the observations ended in 1988). The graphs were updates of a pair I had constructed towards the end of the 1970s at the suggestion of the late Gerry Robinson, so the origin of the book was even earlier. Upon seeing the new graphs, I initiated a search to find out why the changes in two completely different quantities, separated by thousands of miles, should be linked, a search that involved the oceanography and climatology of the North Atlantic, ecology, limnology (the study of lakes), and computer modelling. After the results were published, it became clear to me that there was a popular interest in the topic, for short articles appeared in newspapers and magazines. I was also occasionally asked for radio and television interviews. Even so, the idea of writing a book about the unusual connection didn't crystallise until a visit to the Devon town of Totnes a few years later.

The ancient borough of Totnes, with a prominent position above the River Dart, is one of Devon's gems, full of both colour and character that stems from a rich cultural, historical, and archaeological heritage. It is the second oldest borough in England. From the quay, Fore Street rises up past many fine examples of 16th and 17th century merchants' houses into the centre of the town, passing underneath the East Gate Arch—a splendid Tudor structure. It was in a bookshop on Fore Street, just up from the arch that I developed a plan of the book's layout. Writing the book took rather longer than I anticipated and had the assistance of a number of people.

My special thanks must go to John Stephens who has helped with the analysis of the data used and has assembled the Gulf Stream positions. He has also commented on all the text and prepared many of the figures. I have had discussions with Tony Stebbing throughout the preparation of the manuscript, and he has read substantial portions of the text. The plankton data were supplied by the Sir Alister Hardy Foundation for Ocean Science, some by the late Harry Hunt. The late David O'Leary and Peter Tallack saw the potential of the book and provided advice at an early stage. Mick Jordan has read the whole of the text, and parts have been read by Glen George, Avijit Gangopadhyay, Alistair Lindley and Tony John. Latha Menon and her colleagues from Oxford University Press have been very helpful in seeing the book through to completion. Finally, I want to thank my wife, Venetia, who has also read all the text and has helped with the editing.

Arnold Taylor
Plymouth Marine Laboratory
and
University of Plymouth

1

────── ❧ ──────

A 4000-MILE
PLANKTON RIDDLE

The Danes have noticed that when the winter in Denmark was severe, as we perceive it, the winter in Greenland in its manner was mild, and conversely.

(Hans Egede Saabye, Danish missionary in Greenland, 1745)[1]

When the London Bridge, which had been built in Victorian times, became inadequate for the increase in traffic and began sinking into the Thames, it was sold off and replaced. The old bridge was dismantled and shipped out to a specially constructed site, Lake Havasu, in the desert of Arizona. At the same time that its granite stones were being re-erected in the desert, a quarter of a million miles away a small vehicle was picking its way through another rocky landscape: Apollo 15 astronauts Dave Scott and Jim Irwin were exploring the surface of the Moon in their Lunar rover. In doing so they may have missed an important discovery, for recent observations suggest that soil only metres away from the landing site is coated in water. One of the experiments carried out by Dave Scott was simultaneously dropping a falcon's feather and his geological hammer. Both hit the Moon's surface at the same time confirming Galileo's prediction that this is what would happen in the absence of air resistance—just as well, as the astronauts' ability to get home depended on Galileo being right. Each of these events took place in the year, 1971. In that same year, Mount Etna began spewing forth molten lava in its first major eruption for 20 years, the beginning of a series of 13 such eruptions over the next quarter of a century. This was also the year that East Pakistan became the separate state of Bangladesh.

The year 1971 has another minor claim to distinction, albeit one that is almost unknown. During 1971 the Gulf Stream, whose waters are considered to be so important in ameliorating the UK and west European climate, was further south than it had been at any time in the last half-century. This was not an isolated happening but the extreme of a southward movement that occupied most of a decade. Although observational records only allow the position to be determined over the past few decades, calculations with computer models suggest that this southerly shift might have been unique during almost the last 200 years. This was, therefore, a landmark event. Was it accompanied by significant weather events over the North Atlantic region?

There were, indeed, major storms in the surrounding areas. On 4 March a snowstorm in Montreal killed 17 people and dumped 470 mm of snow on the city. The 110 km/h winds produced second-storey drifts, snapped power poles and felled cables. In total, the city hauled away half a million lorry-loads of snow. In May, heavy rains caused a sinkhole 600 m wide and 30 m deep to appear in a residential area in Quebec. The mudslide killed 31 people and swallowed up homes, a bus and several cars. During August, Hurricane Beth brought punishing winds and 300 mm of rain to the eastern mainland of Nova Scotia. Across the ocean in the UK, 58 mm of rain fell at Walthamstow on 5 August and violent winds occurred in London (January), Newhaven (December) and Norfolk (August). It is tempting to suggest that some or all of these events were associated with the southerly Gulf Stream but, in reality, the likelihood still remains that they represent no more than normal weather variability.

There was one other event, however. During 1971 and adjacent years, there were fewer zooplankton in many of the seas around the British Isles than at any time during the half century over which they had been observed.

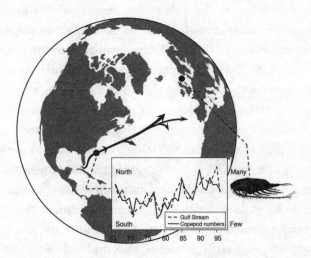

North

Many

South

Few

- - Gulf Stream
—— Copepod numbers

65   70   75   80   85   90   95

FIG 1.1 A climatic connection spanning the North Atlantic Ocean. A considerably enlarged image of a typical copepod is included (the graphs are in standardised units).

Zooplankton are microscopic animals of the open sea which float and drift with the flow of the tides and ocean currents. They are at the base of most of the food chains in the sea and their larger Southern Ocean cousins, krill, are the main food of whales. Subsequently, between 1971 and 1995, the Gulf Stream moved erratically to reach the furthest north it had been observed. This change was accompanied by a similar rise in the numbers of zooplankton, with the animals closely shadowing the Gulf Stream variations. Figure 1.1 summarises this connection, showing the vast separation of the two sets of observations, and how closely the numbers of copepods (the main components of the zooplankton) and the position of the Gulf Stream at the US coast have tracked one another over three decades.[2]

The observation represents a 6000-km link between a vast ocean current and some of the tiniest creatures in the ocean. The animals swarmed more densely in years when the current flowed closer to the US coast than when it flowed further southward. What can such a connection mean? At its root is the atmosphere over the North Atlantic. There is much variability in the ocean over this distance and, because the North Sea is so much shallower than the ocean, the ocean current flows around its entrance rather than penetrating it. The link has arisen because the weather patterns in the two regions are coupled.

This long-distance association is not solely the province of the North Atlantic, for these kinds of linkages, which are called *teleconnections* by climatologists, are known from around the globe. An example very similar to that in Fig. 1.1 is the discovery in the 1990s that the temperature of the equatorial Pacific Ocean can be used to predict the yields of maize far across the world in Zimbabwe. Again it is the atmosphere that is supplying the coupling. Exactly what aspects of the weather are linked together across the North Atlantic is defined by the nature of the two connected observations, the ocean current and the plankton.

The Gulf Stream is a massive current of warm water which sweeps up from the Gulf of Mexico to Georgia before swinging out across the Atlantic and becoming the more diffuse North Atlantic Drift. It carries, as heat, the energy of about 20 million power stations, and this is part of the reason that the climate of western Europe is so much less severe than either the equally northerly Labrador or Novosibirsk in Siberia. The path of the Gulf Stream varies by many tens of miles from week to week and from month to month. Where it is at any time affects shipping, fisheries, and the weather along the Atlantic seaboard of the USA.

In April 1775, as the first shots were being fired at Lexington in the American War of Independence, Benjamin Franklin was returning to Philadelphia from England, following an unsuccessful attempt to prevent the conflict. While crossing the Atlantic he made a series of regular recordings of the ocean temperature.[3] (Fig. 1.2) These data are the first published observations of the position of the Gulf Stream. They also showed for the first time that the northern edge of the current is marked by a sharp fall in temperature. This is also the case at much greater depths, so that the warm current is pressed up against a wall of cold water, now called the 'North Wall'.

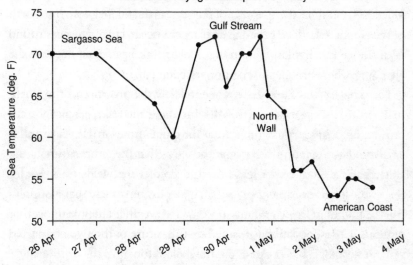

FIG I.2 Franklin's ocean temperatures.

Based on observations of the Gulf Stream's north wall, the US Navy, the US National Oceanic and Atmospheric Administration, and other organisations have published charts since the 1960s showing the path of the current each week as it runs along and then leaves the US coast. These charts are an aid to all kinds of shipping. Realising that they provide a valuable historical record of the current, I devised a way of estimating a mean position of the current which would show whether it was more northerly or more southerly than normal in any year. The procedure was to read from each chart the latitude of the north wall of the Stream at intervals of a few degrees of longitude between Florida (79°W) and Nova Scotia (65°W) and then use a mathematical technique called *principal components analysis* to calculate a weighted average of these latitudes.[4] This process produces an average position, while making allowance for the fact that the current fluctuates more wildly as it moves downstream. The result, called the Gulf Stream north wall (GSNW) index, reveals slow drifts north and south over a range of about 200 km on timescales of years to decades. (There are now other versions of the index based on different sets of data which are generally in agreement with the original index.[4]) Although in the context of the several thousand kilometre-width of the North Atlantic Ocean the displacements are small, they are tied in with the wider circulation[5] and provide a measure of how the ocean is changing. These are the observations used in Fig. 1.1.

But what of the number of copepods? These microscopic animals, making up a substantial part of the zooplankton, are a group of small crustaceans. All are less than 1 cm long and most of them are much smaller. They are found everywhere in the seas and in nearly every fresh-water habitat, even under leaf falls in damp forests, in damp moss, or the water-filled recesses of plants such as pitcher plants. Some of them use the bodies of living animals as homes from which they come and go without detriment to their host, while others are parasites on the gills, skin, or flesh of fish. Copepods are the major grazers of the drifting algae in the ocean, the phytoplankton, whose photosynthesis is continuously absorbing carbon dioxide from the earth's atmosphere and returning oxygen, and they constitute the biggest source of protein in the oceans. The zooplankton observations from which the copepod numbers in the figure were drawn have been collected because of the vision of one man, Sir Alister Hardy.

After serving during World War I, first in the Northern Cyclist Battalion and then as a camouflage officer in the Royal Engineers, Hardy had gone on to investigate the marine biology of the North Sea, often on trawlers and drifters, working on the life history of the herring and its food the plankton. He found that plankton (the collective term for both phytoplankton and zooplankton) gathered into patches, and these affected the fishing. Sometimes when catches were poor the skippers would attribute it to *Dutchman's baccy juice* or a body of water coloured by phytoplankton. In order to investigate this patchiness, Hardy devised what he called the *continuous plankton recorder* (CPR), a torpedo-shaped machine which can be towed behind any ordinary ship at full speed. It automatically samples the plankton near the sea-surface mile by mile as it goes along.

The plankton is sieved out from the water stream by a continuously moving banding of silk gauze, which is slowly wound across the opening and then wound on, rather like the film in an old-fashioned camera, into a storage tank of preservative fluid (formalin). At the end of a run of perhaps two or three hundred miles the spool is taken out from the machine and unwound section by section so that the plankton may be examined, identified and recorded mile by mile. It therefore records a continuous line of observations right across the sea to reveal just how patchy different kinds of plankton can be.[6]

Hardy developed the device for use on a two-year expedition in the Antarctic on Captain Scott's famous old barque-rigged sailing ship, the *Discovery*. (Scott had wished to use the *Discovery* for his final expedition of 1912 but was unable to do so, and sailed in the *Terra Nova* instead.) The expedition's stated purpose was 'the preservation of the whaling industry'. Even so, in his book on the expedition, *Great Waters*,[7] Hardy makes clear his distaste for whale-hunting. The book reproduces many watercolours he painted during the expedition, one of which shows an active day at the whaling station at Grytviken. This painting contains a striking image of their research ship anchored in diluted whale blood. His text reinforces the picture:

It is a barbarous business. These creatures are mammals like ourselves, and I have no doubt that they feel pain. It is amazing that they can put up such a long fight as they sometimes do, after an explosion has occurred in their insides and when every effort they make to get away must increase the agony of the cable pulling in the wound. Perhaps as often as not the whale is killed almost at once, as it should be if the explosive shell reaches the proper target: the main thoracic cavity; but as we have seen it may not. It would never be allowed if it took place on land. Think what an outcry there would be if we hunted elephants with explosive harpoons fired from the cannon of a tank and then played the wounded beasts upon a line!... Of course the whales are huge and the harpoon is relatively small compared to their bulk; it has been pointed out that the size of the explosive head in relation to the size of the large whale is no more, or even less, than that of the service bullet in relation to the body of a man. But, of course, a steel bullet is different; the idea of shooting human enemies with explosive, barbed bullets on lines does not bear thinking of. It is indeed a paradox: I have already emphasised the kind-heartedness of the whaling folk to their fellows. It is amazing how blind and unfeeling man, the carnivorous hunter, can be.

Upon his return, Hardy's vision was to use his plankton recorders to develop a marine survey that would tell us something of the seas as meteorological measurements do for the air. He began having his plankton recorder towed behind commercial ships on routine sea crossings of the North Sea from the early 1930s. Gradually, this programme developed into a regular monthly survey covering the North Sea and a large part of the North Atlantic.[8] These observations have subsequently been used for ecological studies and to form a background for fisheries and research into the impacts of climatic fluctuations. As Bob Dickson of the UK CEFAS (UK Centre for Environment, Fisheries and Aquaculture) Laboratory has pointed out, the seven decades of publications generated by the CPR survey provide a good illustration of how scientific understanding grows with time during the lifetime of a monitoring programme. Initially, the publications were limited to descriptions of what occurs where, but gradually the data became sufficient to contribute to the biology of the plankton, then successively to reveal seasonal changes and dynamics, and eventually year-to-year changes and ecosystem trends. Over the decades, the value of the time-series grew steadily in terms of the breadth of understanding that could be mined from it.[9]

The continuous plankton recorder has subsequently become one of the most successful scientific instruments of all time. Recorders have now been towed for 5½ million miles, and one has even been on view in the Science Museum in London. Since the first tow, more than 200, mainly merchant, vessels have towed the device in a voluntary capacity during their routine sea crossings. These vessels have ranged in size from the *Cotes du Manche*, a French research vessel of 230 tonnes gross, which towed in the eastern English Channel, to the ARCO (*Polar*) *Independence*, which is over 1000 times heavier. Without the help of all these 'ships of opportunity' the research could not have been economically supported.

The CPR survey is also unique in maintaining a sampling system using standard methods for more than 50 years. This consistency of approach has produced long series of observations that can serve as measures of environmental change in the North Atlantic and adjacent seas. There is no other single instrument that has given us so much information on life in the oceans over such a long period of time.

No other observational programme in the world monitors the seas with the geographical coverage of the continuous plankton recorder survey. Other observational programmes exist in which data have been collected at a single point: for nearly half a century the Panulirus station has been monitored in the Sargasso Sea close to Bermuda; the site of ocean weather station PAPA in the eastern North Pacific Ocean has been continuously monitored over several decades; and further south in the Pacific the HOTS site has been maintained near Hawaii. While such programmes as these reveal what is happening at different depths, they give no indication of how changes may be spread geographically. Further, if the direction of water flow through the site varies, the region actually sampled is not really known. On a smaller scale than the CPR Survey, Wolf Greve and colleagues have collected regular samples from an area around Heligoland in the southern North Sea. One survey that does extend outwards over a large region is the California Cooperative Oceanic Fisheries Investigations (CalCOFI) programme, which has operated several times a year off the shores of California. This survey of larval fish began when sardine populations collapsed in the 1940s.

In recent years, surveys using the continuous plankton recorder have been set up around the world. Monitoring of the plankton at the eastern seaboard of the USA, covering the area of the Grand Banks and the Labrador Current, has been operating since the 1970s at the North east Fisheries Science Center in Narraganset, and the Finnish Institute of Marine Research runs a sister survey in the Baltic. Further afield, the Australian Antarctic Division and the National Institute of Polar Research in Tokyo have been running a Southern Ocean CPR Survey around Antarctica. The National Institute of Oceanography at Goa, India is establishing its own Survey. In 2000, sampling in the Pacific Ocean began with tows from Valdez, in Alaska, to California and another tow from Vancouver, Canada to Hokkaido Island, Japan. Extending all the way across the Pacific Ocean, the latter is the longest ever CPR route.

Hardy himself moved on from the Survey to eventually become Professor of Zoology at Oxford University. In the course of his later career, he experimented with sampling flying insects from moving trains, with a view to developing a similar sampling programme to that in the sea. However, this programme required train journeys with a long enough stretch of tunnel- and bridge-free track that a kite could be flown from the train, and so it was never really feasible. He also championed the controversial idea that humans developed and lost their hair because they lived close to water. After retirement, Hardy set up a Religious Experiences Research Unit, in converted stables, with the intention of conducting a scientific study of religion and developing a 'natural history' of spiritual experience. This Unit, which has been collecting all kinds of personal experiences, has carried on following his death, just like the Continuous Plankton Recorder Survey did.

Alister Hardy's major legacy, the CPR Survey of the North Atlantic, generated the numbers of copepods used in the opening illustration (Fig. 1.1). The observations were made in the middle of the North Sea, off Denmark and near the mouth of the Baltic. But the link with the position of the Gulf Stream is not restricted to this location; data from the CPR Survey shows it happening with copepods over 500 miles further north in two areas, one around the Shetland Isles and the other, over 400 miles away, off the coast of Norway (Fig. 1.3), and it has been seen even further away in the eastern North Atlantic.[10]

While the link shows most clearly when all the copepods are pooled together, it appears in varying degrees in individual species. There are some differences because the animals vary in their behaviour, but many features of the response to changing weather are common to most species. Two species from the North Sea exhibiting the connection particularly clearly are shown in Fig. 1.4. When species are grouped together,

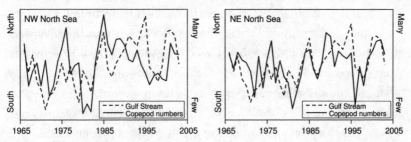

FIG 1.3 Annual changes of some zooplankton populations in the North Sea compared with the position of the Gulf Stream (the graphs are in standardised units).

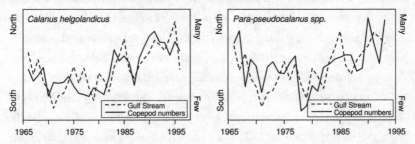

FIG 1.4 Annual changes of two species of zooplankton in the North Sea compared with the position of the Gulf Stream (the graphs are in standardised units).

the individual responses tend to cancel out revealing the underlying link, and this is what happens when only the total number of copepods is used.[2]

The link with the Gulf Stream is, therefore, spread across many species of zooplankton and locations, and not confined to a few dominant species or a few specific regions. The comparisons shown are quite remarkable when the problems in taking the samples and assembling the data are considered. Plankton in the sea is inherently patchy because the animals tend to congregate together. Consequently, it is very easy for a sampling route to miss several intense patches or find an unusually large one. Further, the copepods considered are some of the smaller species and these are less easy to collect and identify than larger ones. Finally, the Gulf Stream fluctuates continuously, meandering back and forth in a serpentine manner, and these variations were averaged out for the graphs. Is it possible that similarities in each of the three figures are no more than a statistical fluke? Gottfried Wilhelm Leibniz, the man who developed calculus independently of Isaac Newton (much to Newton's chagrin), has succinctly summarised this kind of self-delusion: 'credulity fills in the rough outlines shaped by accident'.[11]

The answer to this question is obtained by calculating some measure of how similar each pair of graphs in the three figures are and then computing how often this occurs in randomly generated graphs. Extending as they do over different regions of the North Sea and north-east Atlantic, across a range of species, and over a time period of over three decades, it is perhaps not surprising that statistical analysis shows that relationships like these are very unlikely to be just due to chance. Many tests have shown that these kinds of patterns only rarely happen in random data.[12]

The plankton in the seas around the British Isles are not alone in tracking the north–south movements of the Gulf Stream and it is not only the CPR Survey that has shown up the effect. A linkage with the Gulf Stream has been reported in zooplankton collected off the Northumberland coast by the Dove Marine Laboratory, in zooplankton in the lakes of the English Lake District, and in wild plants in the English Cotswolds and also for the concentration of nitrate during winter in two lakes in SW Ireland. A group from the Institute for Marine Research in Bergen has found that the position of the Gulf Stream can be used to forecast temperatures in the Barents Sea for use in fisheries management.[13] Some of these observations will be discussed in later chapters. The long-distance connection is therefore widespread.

These pan-oceanic associations have implications sufficiently important to merit an attempt to understand them. Firstly, they demonstrate that biological abundance in the sea and elsewhere is determined in any year by climatic factors that operate over many thousands of kilometres. Some of these factors are sufficiently subtle that they reveal themselves most clearly in the ecosystems. Secondly, as a consequence of this, many of the changes in biological abundance between years are not predominantly the result of complex interactions within the ecosystem, or indirect effects of fishing pressure on fish stocks. The intricate network of organisms eating and being eaten by other members of the community, and the increased survival of prey species as the fish that consume them are removed, each have the potential to generate oscillatory fluctuations in communities. It would appear that fluctuations such as these are merely overprinted on a trend that is directed by climatic factors, weather events that are connected across the ocean. Thirdly, the pan-oceanic links can reveal some of the climatic factors controlling the carbon cycle which will be needed to predict its future behaviour. Tracing how the link has operated leads naturally into all these topics.

The long-distance connection is tied into large-scale couplings of the atmosphere across the North Atlantic Ocean. For at least a third of a century, the weather patterns experienced by copepods in the North Sea must have been connected in some way with those driving the Gulf Stream 4000 miles away. The shallow waters in the middle of the North Sea are too far from the North Atlantic Ocean for appreciable effects of the ocean currents to be felt, and so any direct linkage between the Gulf Stream and the plankton of the North Sea must operate through the atmosphere. Further, the lakes in England and Ireland are certainly not subject to ocean currents. Tiny zooplankton are much more likely to be responding to the weather patterns they, their food, or their predators are experiencing. There has therefore been some form of coupling between atmospheric fluctuations along the western and eastern shores of the North Atlantic. The quotation at the beginning of the chapter from the missionary, Hans Egede Saabye, is a very early description of this happening.

Great swayings of the atmosphere, which span a whole ocean, are a common feature of the world's weather. The best known of these is that associated with the El Niño cycles of the equatorial Pacific Ocean, in which there is a seesaw with atmospheric pressure rising (or falling) over Darwin, Australia, at the same time as a fall (or rise) happens thousands of miles away in Tahiti. Even though the North Atlantic long-distance connection is a relatively minor fluctuation by comparison, it shares one feature in common with its powerful cousin: each in some way involves the ocean changing in synchrony with the atmosphere above it. Oceans store a much greater quantity of energy than the atmosphere, and the upper ocean in contact with the atmosphere holds approximately 30 times as much heat as the air above it. Thus for a given change in the heat content, the temperature change in the atmosphere will be around 30 times greater than in the ocean. Small changes in the energy content of the ocean could therefore have considerable effects on the climate of the region. Consequently, the large heat storage of the oceans often acts to control atmospheric changes, and the two media are frequently engaged in an intricate dance. This is especially so in the equatorial Pacific Ocean where the continuous interchange between air and sea is the source of the El Niño phenomenon. Such dances make widespread footprints in the living world, as the Gulf Stream connection shows. Many of the impacts of climate change will be determined by the way the warming manipulates these interplays.

The scientific community now agrees that the world will experience a steady warming over the course of the coming century, and the first signs of this are already appearing. Located in the interior of Alaska on the banks of the Tanana River, Nenana is a town that has long been a traditional gathering place for hunting, fishing, trade, meetings, and celebrations by the Athabascan Indians. The little river community is now made up of miners, trappers, homesteaders, Alaskan natives, shopkeepers, boat crews, and workers on the railroad that passes through the town. Although remote and isolated, Nenana has one claim to fame, an annual contest started in the early years of the last century by bored Alaskan railroad workers which continues to this day under the name of the Nenana Ice Classic. The organisers erect a wooden tripod on the frozen Tanana River, and take bets on when it will fall through the melting ice in the spring. With big prize money at stake—over $300,000 in 2001—the competition attracts wide attention and maintains very precise records, records that now provide a measure of how the local climate has varied through the century. When two researchers at Stanford University, Raphael Segarin and Fiorenza Micheli, looked at these data they found that in recent years sinking has occurred on average 5.5 days earlier than it did in, 1917.[14] This highly unusual data-set provides a tangible example of the rising temperatures which more conventional measurements have also shown occurring in the Arctic.

Although a steady rise in temperature may well occur over half a century, that still does not mean that each year will be warmer than the previous one or that everywhere the temperature will always vary in parallel. While the accumulation of greenhouse gases matters over several decades and weather systems dominate over days and months, over periods of a few years natural variability associated with the large-scale swayings of the atmosphere and ocean are the key. This is because it requires more than 1000 times as much energy to heat one cubic metre of water than the same volume of air, so the oceans are slow to heat up and cool down. Much of the natural variability in surface air temperature between years is due to heat going back and forth between the oceans and the atmosphere. The manner in which global warming impacts on these pan-oceanic swayings will be critical to how the natural variability is affected. For example, it has been suggested that global warming could even lead to a never-ending El Niño. Such an eventuality would dramatically change weather patterns all around the Earth's equator.

Both the steady background warming and any changes in natural variability will leave footprints in the living world. There are many examples of how the warming that has occurred over recent decades has caused populations to gradually spread northward. When walking with my dog along a local beach in SW England, I used to see little egrets close to the water's edge. They are unmistakeable, being pure white birds with long necks, black dagger-like bills and long, black legs that end in bizarre yellow feet. These inhabitants of south European marshland were formerly classed as a rarity in Britain, but are now common and conspicuous year-round residents on many south coast estuaries and coastal waterways. But it is not just birds travelling this road.

As Europe has warmed by 0.8°C over the past few decades, the ranges of many butterflies drifted north in approximate synchrony with the bands of fixed temperature (isotherms). When a group led by Camille Parmesan of the University of Texas analysed distribution patterns of 57 non-migratory butterfly species across Europe, they found that about two-thirds of the species had shifted their ranges northward by as much as 240 kilometres, and climate seemed the probable cause.[15] There is evidence that some butterflies are now beginning to overwinter in Ireland, and new populations are popping up in regions such as Finland and Sweden that were previously too cold for comfort.

This gradual northward drift is also happening in the ocean: zooplankton in the CPR Survey are also starting to show some geographical shifts. Gregory Beaugrand and colleagues at the Sir Alister Hardy Foundation for Ocean Science (SAHFOS), Plymouth, UK, have shown that, over the eastern North Atlantic, warm-water species have extended northward by more than 10 degrees of latitude with a corresponding decrease in the number of colder-water species.[16] The zooplankton have been accompanied by fish populations. In recent years fish species appear to have been spreading northward at something like 50–80 km per decade. There have been increasing reports of southern marine species reaching the British Isles. The Environmental Records Centre for Cornwall and the Isles of Scilly who keep records of warm-water fish new to Cornwall have reported that, between 1960 and 2001, there were no new southern species over the first 15 years but thereafter the numbers have increased at an accelerating rate. One particular example is the sharp rise in off-shore sightings of the ocean sunfish. This strange animal, which resembles a fish head without a tail, is one of the largest bony fishes in the world and can weigh up to 1.5 tonnes. Tony Stebbing, Stella Turk, and others have argued that the increasing occurrence of such immigrant species is associated with rising temperatures in the North Atlantic.[17]

Rising temperatures are gradually changing the northern Bering Sea from an arctic to a sub-arctic ecosystem and, in doing so, bringing in the song of the gray whale. The absence of fish in its upper waters has given the region one of the richest seabed ecosystems in the world. In recent decades, temperatures have risen by around 3°C in the northernmost part of the Pacific Ocean and in the Bering Sea. A 2006 paper in the journal *Science*[18] has shown that, as the waters warm, fish are invading and the bottom dwellers are retreating further north, the fish being followed in turn by their predators, gray whales and walruses. This may mean that whales and walruses may disappear from the traditional hunting grounds of native villagers elsewhere in the Bering Sea.

Not all animals benefit from such changes. Although a shifting range suits some species, for others it means moving out of good habitat into fragmented landscapes where they are unable to survive. In North America, the quino checkerspot butterfly has abandoned the southern edge of its range in Mexico. At the start of the 20th century millions of these orange and black butterflies were spread in a swathe across southern California. Now, warmer temperatures cause larvae to hatch early, and their snapdragon host plants to bloom early, but then the plants also dry out and die sooner, starving caterpillars before they are able to wait out the winter in their cocoons. Meanwhile, habitats in suitable regions farther north, where the butterfly thrived previously, have mostly been swallowed in urban sprawl. Martin Warren and his colleagues of Butterfly Conservation, Dorset, UK, have found similar changes to be widespread among British butterflies.[19] Out of 46 species that approach their northern limits in Britain, three-quarters had declined over the past 30 years because habitat loss outweighed any improvements to their environment coming from climate warming. Camille Parmesan has pointed out that this loss of southern habitats presents the ecologist with a tricky problem. If a species advances poleward its southern extremity may or may not follow suit. How do you demonstrate with certainty the absence of a species? This is a question that may increasingly need to be addressed in the future.

The detailed coverage of the CPR Survey reveals that the effects of the warming have not been evenly spread spatially, through time or across the range of species. Its data have provided other evidence of the effects of rising temperatures on the marine environment. One routine, and somewhat subjective, observation included in the survey is recording how green is the sampling silk. The mesh of the silk used to filter the water in the recorders is chosen to catch zooplankton and, as such, is too coarse to catch the tiny phytoplankton reliably, though some do stick to the silk. Even so, the silks vary in greenness according to how much chlorophyll there is in the seawater. As an estimate of this chlorophyll content, the colour of each silk is compared with a standard scale to give a crude measure of how many tiny plants are present. In a recent analysis of these data, members of the CPR Survey, under their director Chris Reid, found that silks from the North Sea and from the Atlantic to the west of the British Isles became greener over the 30 years up to the 1990s.[20] They pointed out that this may be part of a pattern of change which spans the northern hemisphere.

Ranga Myneni and colleagues of Boston University have found from satellite observations that north of 50°N there was a steady increase of greenness from 1981 to 1991, with the rise becoming most marked in the later years.[21] This increase may be because spring in the temperate parts of the northern hemisphere now starts a week earlier than it did in the 1960s. At first sight the increase in CPR greenness seems to be another manifestation of the increase in vegetation observed from satellites, but this is not the complete picture. Further north, in a large band of the ocean south of Iceland, the exact opposite has occurred, with the CPR silks becoming less green. If these changes are attributable to rising temperatures, the response has clearly not been the same in different places. The response must reflect other weather changes than just warming, or else variations in the responses of different ecosystems.

Closer examination of records from the CPR Survey reveals some of the intricacies of how ecosystems respond to the vagaries of the weather. In the 1970s, I briefly shared an office full of pipe smoke with Michael Colebrook at Plymouth Marine Laboratory, UK, who at that time was making some of the earliest attempts to understand the trends in the plankton populations. A major feature of the data up to that time was a steady decline, widespread through many species. In the central North Sea there was as much as a six-fold decline in the numbers of some species between the 1950s and the 1980s. These observations revealed that planktonic response to natural climatic variability can be quite striking. Similar trends to those in the plankton were seen higher up the food chain, extending even as far as the populations of herrings and kittiwakes.[22] This might have been due, at least in part, to some of these animals being dependent on zooplankton for food, but a common response to changing weather patterns cannot be ruled out.

The decline in the plankton was not uniform across the different species, but was accompanied by other fluctuations from year to year which were more individual. Examining the separate trends more closely revealed some signs of how individual species were affected. Species which occurred in the same places tended to follow similar long-term trends. The climatic processes determining where a species occurs also determine how its numbers vary between different years. It remains to be seen how far these kinds of relationships might continue in a changing world.

The downward trend did not continue beyond the 1970s, but neither was it replaced by a steady rise. Instead, as illustrated in Fig. 1.3, many species became more common or scarce as the Gulf Stream moved north or south. Pooling species together makes this underlying tendency more apparent (Figs 1.1 and 1.2). Superimposed on this ocean-wide connection is the replacement of some cold-water species by their warmer-water cousins and some changes in the greenness of the water. Populations have, therefore, been responding both to changing patterns in the global atmospheric circulation and to the gradual climatic warming. Disentangling these various effects in the future will require ongoing observations continuing for years, and is necessary if signs of the health of the planet's ecosystems are to be correctly interpreted in times to come.

But monitoring programmes are apt to come and go. This was brought out in a letter in the scientific journal Nature by C.M. Duarte and colleagues. As they pointed out, policymakers readily begin monitoring programmes in response to worries about ecosystem change, but funding subsequently dries up. To make matters worse the lifetime of a programme is often close to the minimum period needed for it to show up any significant changes. A consequence of this is that '… the continuation of long-term monitoring programmes is often heavily dependent on the personal effort and dedication of individual scientists.'[23]

An example of the importance of individual scientists to monitoring programmes is provided by the discovery of the Antarctic ozone hole. Joe Farman appeared to be one of science's foot soldiers using an old Dobsonmeter wrapped in a quilt to monitor the thickness of the ozone layer above the UK's Halley Base in Antarctica each year by measuring the ultraviolet radiation penetrating the atmosphere. Invented by Gordon Dobson in 1924, the Dobsonmeter was the earliest instrument for measuring atmospheric ozone. Farman's team observed ozone levels dipping sharply in October 1982 and, 1983. However, the US National Research Council reported that rates of ozone loss observed from satellites were much less than those recorded by the 25-year-old instrument—fractions of 1 per cent. After Farman's results were published in *Nature*, the reason for the discrepancy soon became clear. While the satellites had been monitoring ozone levels around the world, NASA's computer software had been automatically dumping the low readings as false.[24] Without Farman's routine trips to the Antarctic, the dangers of the ozone hole might not have been appreciated for years.

During its long history, the CPR Survey illustrates some of the problems faced by any monitoring programme. It was based at the University of Hull until 1950, when it moved to Edinburgh under the administration of the Scottish Marine Biological Association. In 1976, the whole Survey relocated to what has since become the Plymouth Marine Laboratory (PML). Following a funding crisis in 1989, the Survey has operated as an independent organisation, a foundation named after Sir Alister Hardy— the Sir Alister Hardy Foundation for Ocean Science (www.sahfos.org.uk) now situated in a laboratory with splendid views over Plymouth Sound. Initially, its funding was from British government departments, but now a substantial part of the running costs come from a wide spread of foreign sources.

With current concerns over global warming, the major topic of investigation using data from the CPR Survey is how the long-term changes in the plankton relate to climate changes. The 6000-km link with the Gulf Stream which the observations have thrown up highlights many of the important questions about how weather patterns are interconnected and how they impact on the living world. Solving the riddle of this link necessitates trawling through the processes by which the global climate system operates and ecosystems function. The observed connection encompasses the circulation of the oceans, the motions of the atmosphere, and a range of biological interactions. As such, it touches on most aspects of climate change. Analysis of the pan-oceanic connection also runs up against many of the difficulties encountered in interpreting these kinds of data. Thus, investigating the 4000-mile riddle between the plankton and the distant ocean current entails following a thread through the fields of oceanography, meteorology, and ecology, not to mention the behaviour of complex systems.

The first stage in analysing the connection has to be examining what is known about the ocean circulation and the Gulf Stream, and what factors determine its position. Recording and understanding the currents of the ocean has a long history, from before the American War of Independence, through the early years of Arctic exploration and up to the present. One conclusion from this body of work is that the position of the Gulf Stream is determined by atmospheric conditions over the whole of the North Atlantic and is related to the larger ocean circulation.

What happens to the zooplankton in any year is, to a considerable extent, the result of how the weather patterns affect the surface of the sea, in particular, how the sea warms up during spring and summer. The mechanisms involved are essentially the same in lakes, which, as a consequence of their easier accessibility, have provided much of the understanding. On land, the mechanisms are rather different as populations are subject to different weather effects, e.g. rain and frost. The network of interactions in both aquatic and terrestrial ecosystems complicates any predictions of how they will develop in the future, in the same way that complex physical interactions limit detailed weather forecasts to no more than a few days ahead.

Weather patterns remain at the root of why most ecosystems fluctuate, and they are coupled together in a global interlinked network. Since a classic investigation carried out at the Indian Meteorological Department in the early 1900s, it has been known that this global system oscillates in a set of massive seesaws, each spanning an ocean basin. One of these, the North Atlantic Oscillation, is fundamentally involved in the link between the Gulf Stream and the plankton over the other side of the ocean, not to mention many other changes over the region. It may also be possible that changes in the current system of the North Atlantic can directly influence the atmosphere above them and hence affect populations under the weather systems downstream. Such effects have been proposed in both the North Atlantic and North Pacific, and are certainly part of the El Niño phenomenon.

However, it is not easy to identify specific meteorological changes to which the zooplankton are responding and which can be traced back to either the Gulf Stream or the atmospheric patterns driving it. Ecosystems react in subtle ways to changing weather conditions, sometimes altering markedly, following relatively minor changes. Ecosystems may often respond as a whole to several weather variables, and in this way bring out a common signal as other changes cancel out. If so, ecosystems may respond sensitively to future climatic changes, and in ways that are not expected. That is why we need as much long-term biological monitoring as possible, by both professionals and amateurs, monitoring that will often depend on the efforts and dedication of single individuals.

Elucidating the mechanisms by which the numbers of plankton in the North Sea are linked to movements of the Gulf Stream necessitates exploring the fields of oceanography, meteorology, and ecology. This journey begins in the far north during the heroic days of Arctic exploration.

2

THE WAYS OF THE OCEAN

Call me Ishmael. Some years ago—never mind how long precisely—having little or no money in my purse, and nothing particular to interest me on shore, I thought I would sail about a little and see the watery part of the world.

(*Moby Dick* by Herman Melville)

Fridtjof Nansen was a remarkable man. On 12 June 1895 he and fellow explorer Hjalmar Johansen were returning across the Arctic Ocean, in a pair of kayaks lashed together, from their record-breaking trek which had got within 260 miles of the North Pole. They were sailing westwards with a low undulating ice-bound shore to starboard while to port there was no more land in sight. Although they did not know where they were, Nansen had begun to realise that they were off Franz Josef Land. Towards evening the wind dropped and, after several hours they beached at the foot of the ice in order to stretch their limbs and spy out the way ahead. Mooring their vessel, they climbed a nearby hummock to look around. As they stood there weighing up the situation, they suddenly saw the kayaks starting to drift off. They raced back to the shore as fast as they could, but it was too late: the catamaran had broken its moorings. It was already some way out to sea and rapidly receding. Giving Johansen his watch, Nansen took off some of his clothes as quickly as he could, jumped into the freezing water, and started after the kayaks.

'There went all our possessions,' Johansen recorded, 'Food, clothing, ammunition...our lives depended on retrieving the kayaks. I could not keep still, paced back and forth...could do absolutely nothing whatso-ever, watched Nansen, who rested now and then by swimming on his back; was afraid that he would get cramp and sink before my eyes.'

Nansen wrote in his diary...

The water was icy cold and it was exhausting to swim with clothes on. The kay-aks drifted further and further away. It seemed more doubtful whether I would manage it. But there drifted all our hope, and whether I stiffened and sank here, or returned without the kayaks, the result seemed much the same. So I pressed on with all my strength, and when I became too tired, I turned round and swam on my back...But when I turned round again and found that the kayaks were closer my courage rose, and I carried on with renewed strength. Little by little, however, I felt my limbs become stiffer and unfeeling. I understood that I could not manage much more, but now it was not all that far, and if only I could hold out we were saved. So I forced myself on...At long last, I could stretch out my hand and grasp the ski that lay aft across the kayaks, and haul myself alongside, and we were saved.[1]

Nansen and Johansen's sledge and catamaran journey lasted over a year before they reached a settlement, including a winter spent in a hut surviving largely on polar-bear meat. The trek had begun from Nansen's vessel, the *Fram*, which had been stuck fast in the Arctic ice pack 500 kilometres north of Russia since the winter of 1893/4. This was not due to poor seamanship, but a deliberate attempt, using a specially designed vessel, to undertake scientific study of the Arctic during the inhospita-ble winter. The ice voyage was also used to give a northerly launching point for Nansen and Johansen's attempt to reach the North Pole. Ironi-cally, when they were reunited with the *Fram* at the end of the expedition, they found that, due to the drifting of the ice, they had got only 19 miles further north than if they had stayed on the ship.

But travelling so far north was not Nansen's only achievement dur-ing the expedition. In September 1893 the *Fram* was frozen into the ice, slowly drifting in a westerly direction as the ice was pushed by the prevailing near-surface ocean current. He observed that the drift did not follow the wind, as might be expected, but usually moved at an angle of 20–40° to the right of the wind direction. Nansen rightly discerned that this was due to the rotation of the Earth. He further surmised that, as the depth increased, each successive layer of water, moving over the next like a wind, would produce increasing deflection to the right until, at a certain depth, the movement would be *against* the current at the surface. On his return home to Norway, he discussed this phenomenon with a Swedish physicist, Walfrid Ekman. In a letter to Nansen on 14 November 1901, Ekman developed a theory to explain this process, in which he demonstrated mathematically that the effect of the wind on the surface of the sea produced currents which, in Ekman's own words, 'formed something like a spiral staircase…down towards the depths'. This mathematical model has subsequently come to be known as the 'Ekman spiral'.

These deflections of the currents occur because the water is moving on the surface of a rotating body. The water moves as if there were an extra force, the Coriolis force, acting on the ocean currents, a force which arises because the water is trying to follow a straight line in space, as required by Newton's Laws of Motion, while remaining on the surface of the earth. To illustrate how this happens, consider a circular disc of paper which is rotating anticlockwise on a turntable. If a pen is placed at the edge of the circle and moved along a straight line towards the centre, it will trace a line on the paper that is a curve, the pen acting as if it were continuously being pulled off course to the right of its path (Fig. 2.1). The same is true for any straight movement of the pen. Thus, for a disc spinning anticlockwise, a body moving on it will experience an apparent force to the right. The Coriolis force is named after the 19th century French scientist who first described it. An anticlockwise spin like this is experienced by objects moving on the northern hemisphere of the Earth. In the southern hemisphere, the Earth as seen from in space above the South Pole appears to be rotating clockwise, and so moving objects in this hemisphere experience an apparent force to the left.

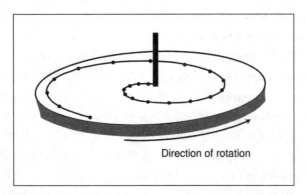

Direction of rotation

FIG 2.1 The line made by a pen moving in a straight line over a rotating disc.

Ekman's theory was a seminal formulation. Appearing at the time when oceanography was turning from an empirical to a mathematical science, it became a fundamental principle of the behaviour of all fluids and gases, and is one of the cornerstones of modern oceanography and meteorology. As a result of this, and his other observational work, Nansen is recognised as one of the founders of oceanography and fluid dynamics. But even these weren't the extent of Nansen's achievements. Before embarking on his Arctic explorations, as a research biologist, he had been one of the developers of a new understanding of how nervous systems function, and after his adventures in the far north, he became a distinguished diplomat. At the end of the 1914–18 war he organised the repatriation of prisoners of war and the supply of aid for Russian refugees, work whose importance was recognised by the award of a Nobel Peace Prize in 1922.

The impact of the Earth's rotation on the ocean is much more wide-spread than the local influence of the wind on the sea surface; it is responsible for the existence of the world's major ocean currents of which the Gulf Stream is perhaps the most well known. The Gulf Stream is a current with a long history whose scientific study goes back two centuries and began with another remarkable man.

On a thundery day in June 1752, Benjamin Franklin, who was already a successful editor and publisher, carried out a legendary experiment in a field near Philadelphia. Flying a kite high up to the thunderclouds he saw a small spark jump from the kite, and in so doing discovered electricity in the clouds and confirmed his hunch that lightning was electrical. Little did he realise how perilously close he had come to electrocuting himself; at least one later attempt to repeat the experiment ended in tragedy. From these early experiments lightning rods earthed to the ground became a standard protection from lightning strikes, and Franklin was established as a great scientist. One group who particularly benefited from Franklin's discoveries were church bell ringers. In medieval Europe, it was common for church bells to be rung during thunderstorms in a vain attempt to break up thunderclouds and keep lightning from striking tall church spires. Ringing damp ropes in wet belfries cost the lives of quite a few ringers. Between 1753 and 1786, lightning strikes down bell ropes killed 103 French bell ringers.

By 1769 Franklin had moved into government and was Postmaster General of the American colony when he came up against a completely different scientific problem. He was asked to look into the reason why mail packets took two weeks longer to cross the Atlantic than did other merchant ships. The explanation for this discrepancy, the presence of a major ocean current, was given to him in a conversation with his cousin, Timothy Folger, a Nantucket sea captain. Folger told him whalers knew that the current, now known as the Gulf Stream, could severely hinder the progress of ships endeavouring to sail against it.

We are all well acquainted with the Stream because in our pursuit of whales, which keep to the sides of it but are not met within it, we run along the sides of it and frequently cross it to change our side, and in crossing it have sometimes met and spoke with those packets who were in the middle of it and stemming it. We have informed them that they were stemming a current that was against them to the value of three miles an hour and advised them to cross it, but they were too wise to be counselled by simple American fishermen![2]

Franklin determined to publicise this current for the benefit of seamen. With Folger's guidance, he assembled the vast experience of the

New England whaling skippers to produce the first chart of the Gulf Stream which was engraved and printed by the General Post Office, and accompanied by 'written directions whereby ships bound from the Banks of Newfoundland to New York may avoid the said stream, and yet be free of danger from the banks and shoals...' Franklin attributed the Gulf Stream to the piling up of water at the American coast by the trade winds, the water 'running down in a strong current through the islands into the Gulf of Mexico and from thence proceeding along the coasts and banks of Newfoundland where it turns off towards and runs down through the Western Islands'.

Heeding the whalers' reports that the Gulf Stream was a warm current, Franklin also pioneered the use of temperature to determine its position. Starting in 1775, he conceived the idea of using the thermometer as an instrument of navigation, and made series of surface temperature measurements while crossing the Atlantic, beginning with those reproduced in the previous chapter (Fig. 1.2). On his last voyage in 1785 he even attempted to measure temperatures to a depth of 100 ft below the surface, first with a sample bottle, and later with a cask fitted with a valve at each end. Tracing currents by measuring surface temperature was subsequently tried by a number of others, one of whom, Captain Strickland, discovered the north-easterly extension of the Gulf Stream towards Britain and Norway at the turn of the century. Although probably known to the Vikings, this extension had escaped the attention of chart-makers.

The Franklin–Folger chart was the first precise chart of the Gulf Stream; earlier charts, made when chronometers were not available to determine ship drift velocities, showed only rudimentary sketches of the currents. The chart subsequently had an eventful history.[3] Partly because British captains maligned it and partly because of Franklin's role in the confrontation between the American colonies and England, copies became very rare and all were then lost for nearly two centuries. After the War of Independence, Franklin published a revised version in 1786, making reference to the earlier version. However, when in later decades all attempts to locate the original failed, its existence began to be doubted. The chart would have remained lost but for the efforts of Philip Richardson from Woods Hole Oceanographic Institution, Massachusetts.

With its shingle houses, sandy beaches, freshwater ponds, and surrounding forest, Woods Hole is an attractive coastal village on Cape Cod, popular with tourists. Extensive, sheltered bays and harbours mean that its population trebles during the summer months. In the 19th century whaling ships worked out of the village and fertiliser was produced from local fish and bird droppings collected on islands in the Pacific Ocean. It is now a world centre for marine, environmental, and biomedical science out of proportion to its small size, being home to several prestigious institutions. One of these, the Woods Hole Oceanographic Institution (WHOI), established in 1930, is the largest independent oceanographic research laboratory in the USA, and has been the focus of many of the most important investigations on the Gulf Stream. Philip Richardson has carried out some of this work.

In September 1978, Richardson reasoned that a copy of the missing chart might have been saved by the French, because Franklin was envoy to France from 1776 to 1785, and both Franklin and his ideas were highly regarded in France. Within minutes of beginning his search, Richardson found two copies of the chart in the Bibliothéque Nationale in Paris.[3] The chart he found still remains a good summary of the mean path and width of the Gulf Stream and the speeds in its core. The Stream is a large and complex current system that fluctuates energetically in space and time and so even today, the system is difficult to measure and interpret, but the measurements that we have agree with the Franklin–Folger chart.

This was not the end of Richardson's historical investigations, for he went on to provide a rare and tragic glimpse of maritime history in the last days of wooden sailing ships. In December 1883, the US Navy Hydrographic Office began to publish monthly pilot charts showing the positions and drifting motions of abandoned derelict sailing vessels and other dangers to navigation in the North Atlantic. Derelicts that survived more than a few days at sea were usually wooden ships, and the longest surviving of these were often lumber schooners. From the point of view of a

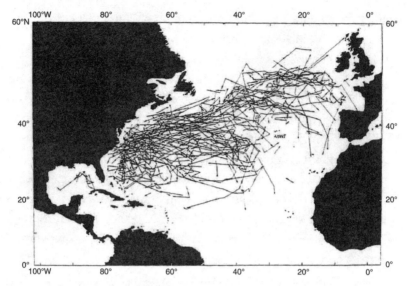

FIG 2.2 The trajectories of drifting derelicts between 1883 and 1902.[4] The general pattern shows up the large-scale ocean circulation over this period. The convolutions of trajectories and their crossing over each other show the variability of the ocean currents with time.

ship's captain, a derelict vessel was a formidable obstruction: a collision with a derelict at night or in fog could damage or sink a ship. In our age of metal ships, it is not generally realised how many derelicts there were, or how long they remained afloat. The Atlantic was literally strewn with ships in various stages of disintegration. Between 1887 and 1893, 1628 derelicts were reported, of which about a third were floating bottom up.[4]

Using the repeated sightings of derelicts identified by name, Richardson was able to produce a chart of the trajectories, which reveals what the circulation and the variability of the ocean currents were at the end of the 19th century. He was then able to compare the chart with another based on satellite tracking of freely drifting buoys. Even in spite of the great variability in the drifts, his chart from the derelicts (Fig. 2.2) looks very much the same as that of the paths of the modern buoys (Fig. 2.3) —the basic surface circulation pattern has not changed. All the drifting

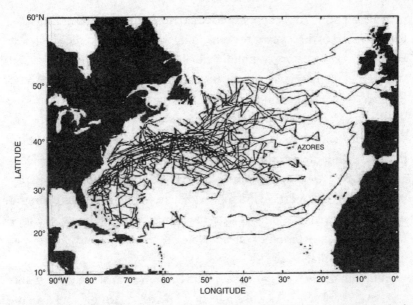

FIG 2.3 The trajectories of freely drifting satellite-tracked buoys between 1971 and 1981.[4]

derelicts followed the path of the Gulf Stream system from the coast of the USA near Florida northward and westward in the direction of the British Isles. Some continued on that way, but others branched off at the Grand Banks off Newfoundland to return in a loop back towards where they started. This branching off is also seen in the drifting buoys. The centre of the large gyre formed by the returning currents, which is located north-east of Florida and the Caribbean, is a large region of the ocean called the Sargasso Sea.

There have been even more unusual tracers of the ocean circulation. During a storm in January, 1992 some containers from a Chinese factory being shipped to the USA were washed into the Pacific Ocean. At least one container broke open, spilling its cargo of 29,000 plastic bath toys —yellow ducks, red beavers, green frogs, and blue turtles. An American oceanographer, Curtis Ebbesmeyer, has been tracking these bath toys ever since. Two-thirds of them floated south through the tropics, landing months later on the shores of Indonesia, Australia, and South America. Over a period of three years, the remaining 10,000 toys travelled east past Japan and Alaska before continuing north towards the Arctic. Many were stranded in the Bering Strait but thousands became frozen in the Arctic pack ice, moving at a mile a day towards the Atlantic Ocean. In 2000, the first toys, now bleached by sun and seawater, reached Greenland and soon appeared on the eastern shores of the USA. Some are expected to follow the Gulf Stream across to the British Isles. The US National Oceanic and Atmospheric Administration has estimated that, driven partly by the wind, the toys travelled approximately 50 per cent faster than the water in the current. Stamped with the words 'The First Years', the toys have subsequently become collectors' items, changing hands for several hundred US dollars each.[5]

Although the two sets of observations a century apart in Figs 2.2 and 2.3 reveal a very similar ocean circulation, there is clearly great variability present in all the drifts. It arises because the currents are constituents of a very dynamic system of surface circulations driven primarily by the winds, winds which fluctuate wildly from day to day and from place to place. Even so, the winds still have preferred directions, about which they fluctuate, and these general preferences are picked up by the surface currents of the ocean. In each of the ocean basins, the trade winds near the equator push water towards the west, while the westerly winds in the mid latitudes push it towards the east, a pattern that results in great current gyres that spin clockwise in the northern hemisphere and anticlockwise in the southern hemisphere. The trajectories of drifting buoys in Fig. 2.3 trace out this gyre as do the derelicts in the 19th century (Fig. 2.2). This gyre is presented schematically in Fig. 2.4. Under this view, the Gulf Stream is not an isolated current; it is the western rim of a giant, spinning lens of water.

The hub of this great circulating wheel of water is a warm, deep volume of slowly moving water, the Sargasso Sea, which is at its thickest in the central area of the gyre of currents. In the 15th and 16th centuries, when the Atlantic was being explored by sailing ships, the Sargasso Sea

FIG 2.4 The current system of the North Atlantic showing the Gulf Stream running along the seaboard of the USA.

was considered frightening and mysterious. A berry-like weed grows here, and because ships were often becalmed in these latitudes, the weed could be seen for days on end, encouraging the superstition that it was a trap for flotsam and could enmesh ships in a solid mat of growth. These were, of course, the days when ocean deeps were believed to house many monsters that could demolish ships or consume sailors. The reality is that the slack currents in the Sargasso Sea allow this weed to grow and float as do the freshwater weeds on a garden pond.

The currents surrounding this Sea are not symmetrical, however: the western rim of this gradually spinning lens is not the same as the eastern. For a start, the centre of the gyre is not in the centre of the ocean, equidistant between Europe and North America but is displaced to the west, into the western Sargasso Sea. To the east of the centre, water flows southward in a broad, barely perceptible drift, moving at only half a knot or less. To the west the same volume of water is forced through a much narrower channel, travelling as fast as 5 knots in some places—this is the Gulf Stream. This asymmetry can be seen in Figs 2.2 and 2.3 where the drifting markers are packed closely together travelling northward along the coastline of the USA, but are sparsely spread on their more eastern southward journey. This east–west difference is not unique, for the same kind of intense *western boundary currents* as the Gulf Stream are found in all the oceans. Other examples are the Agulhas Current, which flows south along the east coast of South Africa and attains current speeds comparable to the Gulf Stream, and the Brazil Current flowing south along the coast of South America. The Pacific Ocean, which occupies almost half the earth's surface, has its own strong current, the gigantic Kuroshio, which flows north along the coast of Japan and Siberia.

As in the North Atlantic, the Kuroshio is part of a swirling vortex of currents comprising most of the northern Pacific Ocean, known as the North Pacific Gyre. At the centre of the Gyre is a relatively stationary Pacific equivalent of the Sargasso Sea, which collects material from the circulation around it. This area accumulates all kinds of flotsam and other debris. In the past this would have steadily degraded, but now, rather than breaking down entirely, modern plastics disintegrate into smaller and smaller pieces, eventually becoming individual polymer molecules which are not easily digested by bacteria and which can attract chemical pollutants. These floating particles sufficiently resemble zooplankton that they may be consumed and enter the food chain. Samples taken from the Gyre have found the mass of plastic present in the water to be several times that of the zooplankton, and have shown that they extend over an area about the size of Texas. The region, informally called The Great Pacific Garbage Patch, is arguably the largest rubbish dump in the world.[6]

The east–west asymmetry in the currents of the world's oceans was long one of the great mysteries about surface currents until in 1946 Henry (Hank) Stommel from WHOI first heard about the problem on a car trip to Rhode Island. As his colleague, Ray Montgomery, later recalled,

Stommel outlined a solution to the problem over a coffee break on the journey. Subsequently, by the insightful simplification of imagining the ocean as a rectangular box with a flat bottom, he showed that the asymmetry happens because on a spherical earth the Coriolis force varies with latitude. This discovery made his name in oceanographic circles and set him off on an illustrious career.[7]

Stommel's solution was achieved by applying a universal physical law, the conservation of angular momentum, to the ocean.[8] The angular momentum of anything that spins is obtained by multiplying together its mass, the square of its radius, and its rate of spin. According to the conservation law, the angular momentum of any object must remain constant unless it is being twisted by some outside force. Thus, when a figure skater pulls in her arms, decreasing her radius, she spins faster so that her angular momentum remains unchanged. However, unlike the ocean, the figure skater is held together as a single entity—she doesn't spill out in all directions.

The figure skater's counterpart in the ocean is a vertical column of water and its angular velocity is represented by its *vorticity*. This is somewhat analogous to the way atmospheric movements are thought of as rotating weather systems progressing over the planet's surface. Vorticity can be thought of as the tendency to rotate, and all seawater has it, even water that is not actually moving in circles, but merely along a curved path. Every water column gets vorticity from two sources that are sometimes in opposition. First, being blown by the wind or meeting a coastline may set a column of water spinning on its axis relative to the rest of the Earth's surface. This is what the trade and westerly winds do to the whole of the North Atlantic Ocean. Vorticity of this kind is called *relative vorticity*. But the Earth itself is rotating all the time. Just by virtue of being on a spinning planet any water column partakes of a *planetary vorticity*, which is anticlockwise in the northern hemisphere and clockwise in the southern. Relative vorticity plus planetary vorticity equals the water column's absolute vorticity. If the depth of the water column remains unchanged, this is what is conserved in the ocean.

Stommel's key insight was to recognise that, because the Earth is almost spherical in shape, the Coriolis force increases with latitude, and as it changes, so does the planetary vorticity. Any person standing at either pole is twisted around over 24 hours and so a column of water here has some planetary vorticity: anticlockwise in the north and clockwise in the south. However, the same person standing on the equator is merely swept around a circle without twisting, and so water columns here have no planetary vorticity. Thus, when a column of water moves from one latitude to another, its planetary vorticity changes. Because its overall absolute vorticity must not change during such a move, its relative vorticity must also change by the same amount, but in the opposite direction. This variation exerts a big effect on the currents of the oceans.

The current gyre of the North Atlantic is driven by the winds and has settled into a steady state in which the force of the winds is balanced by frictional forces. These increase as the current flows faster. Figure 2.4 shows that the currents generated by the winds follow a clockwise gyre and so this is the direction of vorticity imposed by the winds. As the currents brush against the European and American coasts clockwise vorticity is drained from the gyre, thereby balancing the inputs from the winds. However, this balance is achieved in different ways on the east and west sides of the gyre. As water moves south towards the equator it loses anticlockwise planetary vorticity on the east side, so that, to keep its total vorticity the same, it must lose some of its clockwise relative vorticity as well. That leaves less to be done by friction, which explains why the current in this half of the ocean can be a broad, slow drift. But on the western side of the Atlantic, and other oceans, the situation is reversed. As the water moves north along the coast, it gains anticlockwise planetary vorticity and, to balance this, must gain clockwise relative vorticity as well. Movement northwards is aggravating the effects of the winds and the only way for the gyre to get rid of all this clockwise vorticity is to have a lot of friction, which requires a fast current.[8]

This then is why there is a Gulf Stream. It is also why there is the Kuroshio in the North Pacific Ocean, the Brazil Current in the South Atlantic and the Agulhas Current in the Indian Ocean. In each case the strongest current is that moving towards the pole, and the wind patterns dictate that the poleward flow is at the western side of the ocean.

While this is the basic reason for the Gulf Stream, the physical processes that define its path are more intricate. The current runs up the US coast but separates from the coast and flows into the central Atlantic at Cape Hatteras. The physical mechanisms responsible for this separation, which need to take account of what the density of the ocean is at different depths, are so involved that even some of the most advanced models of the ocean circulation have had difficulty reproducing it; frequently, the simulated current goes much too far north before leaving the coast. The mechanisms operating are, of course, especially important for understanding how the point at which separation occurs varies from year to year. Although the details may be subtle and not easy to compute accurately, the underlying processes may still be reproduced in simple models.

In 1969, A.T. Parsons of Bristol University published just such a model showing how the wind stress across the Atlantic operates on the ocean density field to cause the current to separate from the coast. Parsons was one of those scientists who make an important breakthrough early in their careers, then move into a different field of work (in Parson's case, the UK Admiralty Research Laboratory) and consequently never pursue the topic further. In his model, the ocean is represented by two layers: a surface layer and a denser deep layer. The deep layer becomes exposed at the surface along the Gulf Stream's north wall, and the position at which this occurs is determined by the balance of vorticity. Parson's model was subsequently applied to all the world's oceans by George Veronis at Yale University.[9]

More than a decade after Parson's work, Avijit Gangopadhyay (then at the University of Rhode Island), together with some colleagues, decided to test if the simple model was sufficient to account for how the latitude at which separation occurred varied from year to year.[10] To do this, they compared observations of the Gulf Stream from satellites with the predictions of the model. When they did so, they found that, to obtain agreement between the model's results and the observations, they had to make predictions using an average of the wind stress over the previous three years. In reality, their results suggest that the point at which the Gulf Stream leaves the continental shelf is affected by the wind stress off Europe more than six years ago, off the Azores in the centre of the ocean more than four years ago, and off the North American continent several weeks to several months ago.

Gangopadhyay's group suggested the time delays happen as a result of the energy brought across the Atlantic by so-called *Rossby waves* in the different density layers of the ocean. These waves, which are generated from fluctuations in the wind field, are dependent on the oceanic density field and, driven by the earth's rotation, travel to the west. Similar waves occur in the atmosphere where they are even more important. These waves have various lengths and timescales, but it is thought that, in the ocean, the most important of these have length scales of about 100 km and periods of about six months. Moving very slowly (several kilometres per day) beneath the surface waters, they are difficult to observe. At this kind of speed the waves will take several years to cross the Ocean from side to side and this may explain the three-year delay between the wind forcing and the movements of the Gulf Stream.

Subsequently Avijit Gangopadhyay moved to the University of Massachusetts, Dartmouth, based at New Bedford, a town that grew very prosperous in the 19th century out of whale hunting, and is famous because Hermann Melville, the writer of *Moby Dick*, stayed there before sailing on a whaling ship. Oil from sperm whales was valued as a lubricant and was still in use by NASA during the 1960s. Almost all the tapers, lamps, and candles that burned around the globe in Melville's day were derived from whales. He described the town as perhaps the most expensive place to live in, because '..nowhere in all America will you find more patrician-like houses; parks and gardens more opulent than in New Bedford.' Avijit and I have collaborated in adapting a simple model, developed by a group that included Henry Stommel of WHOI, to show that the movements of

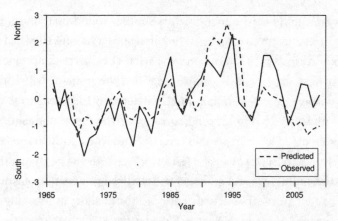

FIG 2.5 Latitude at which the Gulf Stream leaves the US coast as predicted by the Behringer–Regier–Stommel model compared with observations over 40 years (the graphs are in standardised units).

the north wall of the Gulf Stream seen in the last chapter are caused by the same kind of delay processes (Fig. 2.5).[10]

All these results impinge on the long-distance plankton riddle. As the Gulf Stream's position is a delayed response to wind changes operating over about a three-year period, then, if there is to be a climatic linkage tying the changes in the numbers of copepods to the position of the Gulf Stream, it would seem that the biological communities must also show a delayed response to the weather of up to three years. But the copepods spawn a new generation every few weeks and there seems to be no way populations floating in the open sea or ocean can affect their descendents three years hence. The simplest explanation is that the copepods just respond to the immediately recent weather, which is itself a delayed response to the Gulf Stream position. This would mean that the correlation in Fig. 1.1 represents a real cause and effect relationship. But could this be what was actually happening? Answering this question has to be deferred until the climatic and ecological processes involved have been outlined.

South-east of New England, the Gulf Stream's path varies substantially from week to week. Its meanders (or lateral shifts) have a whip-like nature with a wavelength—its distance downstream from one crest to the next—ranging from 200 km to 1000 km. The path can shift north or south by as much as 300 km, which is much more than three times the typical width of the current. Thus, sailing south from Narragansett Bay in the path of the old whalers, the Gulf Stream might be reached in as little as 150 km or in as much as 450 km, depending on which way it has meandered. The current meanders further and further as it moves downstream after leaving the US coast close to Cape Hatteras. These north–south gyrations range from multi-year and seasonal shifts, through variations lasting a month or two, to meanders with periodicities as short as four days on the eastern part of the path close to Florida.

The Gulf Stream therefore snakes about in a tortuous way, but its behaviour can be more complex. Large loops in the path can grow and pinch off to form closed rings of current that may persist for weeks to years before they coalesce back into the Stream. The formation of these rings has close dynamic analogies with the way wind patterns in the upper atmosphere relate to the familiar high and low pressure regions and storms, which appear on weather charts. Rings from the Gulf Stream are the most energetic of these 'storms' or eddies that occur anywhere in the ocean.[11]

Eddies, which are the ocean's equivalent of weather systems, occur throughout the world's oceans. All meandering western boundary currents shed them during their unstable wandering, but other currents do so as well. Seasonal eddies appear in the Somali current off eastern Africa, and eddies are created through instability in the Antarctic Circumpolar Current, which flows all the way around the great Southern Continent. The closed rings of warm or cold water that are constantly spinning off the Gulf Stream are readily visible on satellite images at the ocean surface but eddies may spin the cones of water all the way to the bottom.

Eddies may be the weather systems of the ocean, but compared with atmospheric storms, which last only for weeks, eddies are all remarkably long-lived, persisting in the open ocean for many months. Another difference is size. Cyclones and anticyclones in the atmosphere are more than 1000 kilometres in diameter, but their cousins in the ocean are at least ten times smaller. Mainly, these differences arise because water is so much denser than air: small changes in the ocean produce substantial pressure changes. Associated with the greater density is greater inertia and this causes the eddies to persist. Because they are such compact rotating masses of water, the currents within eddies are strong, flowing at many times the slow background drift of the ocean.

A consequence of the differences is that there are many more eddies around in the ocean than storms in the atmosphere. When Philip Richardson compiled all the available observational data from Gulf Stream rings and other eddies,[12] he estimated that at any given time there were around 1000 eddies drifting around the North Atlantic, spaced on average about 120 miles (200 km) apart and each about 50 miles (80 km) across. The ocean it seems is a mosaic of eddies, and as a result when oceanographers put a group of floats into a current like the Gulf Stream they get a tangle of looping crossing tracks, on which the Stream itself can be barely made out. This chaotic pattern is very apparent in the trajectories of the satellite-tracked buoys (Fig. 2.3), and in the drifting of the 19th century derelicts which were also spun haphazardly by the fluctuating currents (Fig. 2.2).

In the atmosphere the cyclones and anticyclones that make mid-latitude weather so variable are responsible for a major part of the transport of excess heat from the equatorial regions to the poles. This is not the case in the oceans, for oceanic eddies play a much smaller role than the poleward currents. The main reason for this contrast is that the eddies are so much smaller than the atmospheric systems. Atmospheric eddies also travel much faster, crossing the Atlantic in a few days, while a Gulf Stream ring may wander only a few hundred kilometres over a season. While they may not play a large role globally, eddies can still have local effects.

Because it is an almost enclosed sea surrounded by warm lands and is subject to continuous evaporation, the Mediterranean Sea is warmer and saltier than the Atlantic with which it connects. In spite of its water being warmer, the saltiness still makes it quite dense and at intermediate depths, of two-thirds of a mile or so, the Mediterranean Sea is constantly spilling warm, salty water into the Atlantic. Close to the entrance to the Mediterranean the ocean temperature falls steadily with increasing depth, but then approaching 1 km down as this spillage is reached it starts to rise again, and continues to do so until the layer is passed. Some of this spillage is spat out in the form of eddies. In 1984 a team of oceanographers dropped a float into one of these 'Meddies' off Cape St Vincent, Portugal. Although by that time the Meddy was two years old, the researchers managed to track it for another two years, as it drifted nearly 700 miles south. Spinning clockwise every five or six days, it gradually shed its load of two and a half billion tons of excess salt and its excess heat, until it finally exhausted itself and became one with the Atlantic.

One place where eddies may be very important in redistributing heat is off South Africa. The Agulhas Current, flowing south along the East African coast, overshoots south and west of South Africa and turns east with the Circumpolar Current. However, some of the water transported from the Indian Ocean is spun off into eddies that enter the Atlantic and, in addition to a flux of water to the Atlantic, these eddies also carry heat.

Oceanic eddies can also be biologically distinct from the surrounding waters. Eddies originating in warmer waters, warm core eddies, tend to be biologically poor with a limited nutrient supply. By contrast, those coming from cooler waters, cold core eddies, are very productive, as there is commonly a good supply of nutrients from the source region. Caged in by the wall of rotating currents, eddies sometimes contain populations quite distinct from those of the surrounding ocean. Eddies are therefore a source of some of the biological variability in the ocean.

Winds are not the only cause of ocean currents: as shown by the Mediterranean outflow, they can result from differences in density. For example, cooling can make the waters become denser and sink, whereupon less dense water flows in to replace it. Surface motions generated by the action of the winds have been studied for a longer time because wind-driven models of the ocean proved to be easier to formulate. Even so, for almost a century, oceanographers have puzzled over how much of the ocean circulation is driven by the stress of the winds, and how much by geographical variations in density caused by the sun's uneven heating, by the cooling of seawater in high latitudes or by the effects that evaporation and rainfall can have on the concentration of salt in the ocean.

In the 1870s the journal *Nature* carried a series of letters debating these two causes of the ocean circulation.[13] One participant, Dr W. Carpenter, argued that density differences drive ocean currents. He may have been a little pompous, for some of his companions on the HMS *Challenger* had a parrot that they taught to say 'Aha, Dr Carpenter, FRS'. His adversary, James Croll, who was at the time a janitor at the Andersonian College and Museum in Glasgow, propounded the view that the currents are driven by the wind. The correspondence grew heated and personal until finally an anonymous third letter writer (the identity was lost when the files of *Nature* were destroyed during the London Blitz) called for the exchange to be suspended. It is now understood that both processes are equally important. While wind-forcing drives currents close to the ocean's surface, beneath this layer is a much more gradual set of motions that are caused by density differences but these are spread over a thicker body of water.

Changes in seawater density arising from variations in temperature and salinity drive a movement in the deep waters of the ocean called the *thermohaline circulation*. Where the water becomes denser than the deeper layers, it can sink to great depths, although in practice there are only a relatively few regions on the Earth where such sinking has a major impact. *Deep water*, defined as water that sinks to middle levels of the major oceans, is formed only around the northern fringes of the Atlantic Ocean, in the Labrador and East Greenland Seas. *Bottom water*, which constitutes a coldest and densest layer and lies below the deep water, is formed only in limited regions near the coast of Antarctica in the Weddell and Ross Seas. This cooling and sinking drives a massive movement of water connecting up the world's oceans, which is known as the *ocean conveyor belt* (Fig. 2.6).

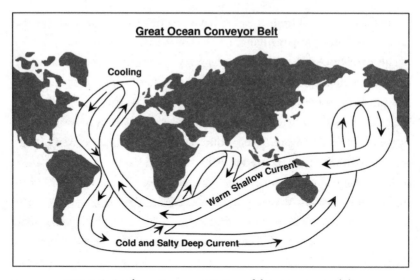

Great Ocean Conveyor Belt

Cooling

Warm Shallow Current

Cold and Salty Deep Current

FIG 2.6 A schematic representation of the *ocean conveyor belt*.

This interpretation of how the world's oceans circulate, which was developed by Wallace Broecker of the Lamont Doherty Laboratory, is probably a simplified representation of a more complex system, but it nevertheless provides a good starting point for appreciating the climatic implications of the thermohaline circulation.[14] Wally Broecker is a chemical oceanographer with an interest in tracing chemical constituents and isotopes around the globe, and whose concerns about global warming go back to the 1970s. The ocean conveyor belt is a continuous system and, as such, does not begin and end anywhere. For convenience, let's suppose that the conveyor belt starts north of Iceland, a hundred miles off the coast of Greenland with, say, a winter's night when the west wind has been screaming off the ice for days. As the surface waters get colder and colder they become denser and sink, probably in very localised regions up to a few tens of kilometres in diameter. The sinking waters fall freely until they reach the bottom, more than a mile and a half down, where they join a pool of deep water that fills the Greenland and Norwegian basins.

This pool of cold water is held back by a dam, an undersea ridge stretching between Greenland and Iceland and Scotland but from time to time the pool spills over it. The water cascades over seafloor lava flows and sediment drifts, falling into the Atlantic abyss and dragging in shallower water as it goes. Reaching the latitude of Newfoundland, the water is joined by similar water from the Labrador Sea, where winters are also severe, but as this water is not quite as dense, it settles at a shallower depth. Near Bermuda the water is also joined by spinning blobs of warm, salty Mediterranean water and all these different waters gradually mingle to become what is known as North Atlantic Deep Water.

This mighty mass of water, equivalent to about eighty River Amazons, creeps slowly over the ocean floor southward across the equator. The flow continues to hug the western slope of the ocean basin through the South Atlantic until it passes the tip of South America, enters the stormy Southern Ocean and meets the Antarctic Circumpolar Current. This sweeps around the frozen continent from west to east, touching every ocean without meeting any land, and transports the equivalent of more like 800 Amazons. Its water is not so cold and is less dense than the frigid waters closer to Antarctica, and so floats upward as part of a strong vertical circulation. In its journey around the south of the world, the water spreads north into all the oceans, mostly at a depth of about half a mile, some into the Indian Ocean but the largest fraction into the Pacific. In each basin the water drifts northward, hugging the western slopes where possible, until it reaches the equator where the mixing of the trade winds brings it into contact with surface waters.

The water now begins the return journey to the North Atlantic, funnelling through the islands of Indonesia and the Philippines, before crossing the Indian Ocean and heading for the southern tip of Africa. Rounding the Cape of Good Hope is no easier for water than it is for ships, and only by becoming detached from the main current in spinning eddies does some of the water manage to struggle into the South Atlantic. One last obstacle still remains: crossing the strong east–west currents set up by the Atlantic trade winds. Giant eddies spin the water north along the

Brazilian and Venezuelan coasts, before dumping the water into the Gulf Stream at its source off Florida. The flow of the Gulf Stream returns the water to its starting point in the northern North Atlantic, thereby completing the circuit, a round trip that has lasted about a thousand years.

In the North Atlantic, the ocean conveyor belt is driven by the production of cold, dense water that sinks and is replaced by warm, surface water flowing up with the Gulf Stream and its extension, the North Atlantic Drift. This circulation is sensitive to milder winters in the northern regions, and to pulses of less dense meltwater into, or close to, the areas where deep water forms at high latitudes.[15] It is therefore affected by precipitation, evaporation, river flows or sea ice formation. If either warming or increased freshwater content of northern waters leads to less cold dense water sinking, this will have ramifications further around the conveyor belt. For this reason, the North Atlantic thermohaline circulation has been referred to as the Achilles' heel of the ocean conveyor belt. Further, the sinking draws massive amounts of heat northwards from the tropics, and so any variation in its intensity will potentially have significant effects on the climates of Europe and the eastern USA. Should the northward flow be greatly reduced, the prevailing westerlies would be blowing over much colder water before reaching Europe. This situation is different from the other major northern hemisphere ocean, for there is no corresponding Achilles' heel in the North Pacific. In this ocean the water is less saline, and hence less dense, with the result that this kind of deep sinking does not occur. Therefore, the North Atlantic is uniquely vulnerable to disturbances of the winter sinking process, a danger that has been revealed by computer models.

At the end of the 1980s, Syukuro Manabe carried out a significant experiment with a climate model developed at the Geophysical Fluid Dynamics Laboratory in Princeton, New Jersey. He allowed the concentration of carbon dioxide to keep increasing at its present rate until, after 140 years, its atmospheric concentration had quadrupled. The Earth's temperature in the model steadily rose and, as it did, so did the amount of water vapour in the atmosphere, whereupon the winds carried much of that water to high latitudes, and it fell as rain and snow. The rivers of the far north, such as the Mackenzie and the Ob, became swollen torrents emptying water into the Arctic Ocean and the Greenland Sea. By the 200th year of the simulation, a dramatic change had occurred: the thermohaline circulation had stopped dead.[16]

Subsequent modelling work has confirmed this sensitivity to additions of freshwater in the far north, although the changes are not as severe as in the first generation of climate models.[17] The thermohaline circulation is susceptible to run-off from the continents, the number of icebergs calved off Greenland and the amount of precipitation from low pressure systems tracking north-eastwards past Iceland into the Norwegian Sea. Small changes in the combined impact of these processes may trigger sudden switches to alternative patterns which carry less warm surface water northward to replace water sinking and returning southwards. Such changes would reduce sea surface temperatures around southern Greenland and Iceland by 5°C or more, with drastic impacts on the climate of Europe, and would completely alter the atmospheric circulation of the northern hemisphere.

All these predictions are more than theoretical possibilities for there is evidence that such changes have happened in the past. About 13,000 years ago at the end of the last ice age, a giant lake, three times the size of the present-day Lake Ontario, burst its dam and flooded down the Hudson River Valley past the site of New York and into the North Atlantic. Lake Iroquois had been formed as the Laurentide Ice Sheet receded. From an analysis of sediments in the region, Jeffrey Donnelly from Woods Hole Oceanographic Institution and colleagues have inferred that the sudden flux of meltwater caused a slowing of the thermohaline circulation and of the heat transport to the North Atlantic. Marine and terrestrial proxy records show a cold event lasting about 300 years, the Intra-Allerød cold period which began at this time.[18]

The circulation may even have faltered in more recent times. Between the mid 16th and mid 19th centuries Europe and all countries bordering the North Atlantic experienced a prolonged cool period, which has come to be known as the Little Ice Age. A striking image from the Little Ice Age is that of the River Thames in London freezing so severely that Frost Fairs could be held on the ice. One Fair in the particularly severe winter of 1683–4 was described by the diarist John Evelyn:[19]

> The Thames before London is still planted with booths in formal streets, and with all sorts of trades and shops furnished and full of commodities of all kinds. Coaches ply to and fro as if in the streets, and there is sliding, bull-baiting, horse and coach races, puppet-plays and other interludes, tippling and other lewd entertainments—so that it all seems to be a bacchanalian triumph, or a carnival upon the water.

In all there were ten winters in the 17th century when the Thames froze solid, and another ten in the following hundred years or so. The last Frost Fair was over a few weeks in January, 1814. Elsewhere in Europe, the same image of bitter winters prevailed, together with periods when cold, wet summers destroyed harvests, and the glaciers of the Alps and Scandinavia advanced. In the worst winters, ice blocked the harbour of Marseilles for weeks and the canals of Venice froze over, as did the rivers Tiber in Italy and Ebro in Spain. In America, the pioneer settlers faced severe winters. However, although overall the Little Ice Age was a very cold period, some decades were much warmer and the climate fluctuated wildly. The mid 1660s were so warm that drought struck Britain in summer, hot weather which encouraged the outbreak of plague in 1665 and produced tinder dry conditions in 1666, helping fan the flames of the Great Fire of London.

The violin-maker Antonio Stradivari lived during the Little Ice Age when the trees of Europe were struggling[9] producing some of the narrowest rings of summer growth seen in the last half-millennium. That may partly explain why a Stradivarius can be worth over a million pounds: in 2003 a tree-ring expert, Henri Grissino-Mayer of the University of Tennessee, and a climate scientist, Lloyd Burckle of Columbia University suggested that the narrow rings made the spruce wood used by Stradivari exceptionally strong and dense with musical qualities that later violin-makers have never been able to achieve.[20]

In 1832 an extensive compilation of observations by James Rennell from the British Admiralty was published posthumously, data which provides a picture of the North Atlantic circulation at the end of the Little Ice Age. His authoritative and exhaustive work[21] delineated and charted the major features of the Gulf Stream. It showed, for the first time, that the breadth of the Stream changes from time to time, doubling or halving even within a few weeks. Rennell's compilation even recorded the formation of giant eddies with cold patches at their centre. While all these observations are in agreement with the modern view, Hubert Lamb, a climatologist at the UK Meteorological Office, pointed out that the currents and sea surface data show the Gulf Stream following a different path from its present course.[22] The waters of the Stream were further south than now and tended to turn away before reaching the coast of Europe. At the same time, the northernmost Atlantic was colder, ice cover was more extensive, and depressions followed more southerly tracks. All this could indicate that some weakening of the ocean conveyor belt had occurred.

The most advanced climate models being used to forecast how the world's climate will evolve as greenhouse gases accumulate in the atmosphere include the dynamics of the ocean circulation. According to the latest report of the Intergovernmental Panel on Climate Change, the consensus of these model predictions is that the thermohaline circulation will weaken in the manner of Manabe's calculations. But even so, the decline is not expected to be large enough to counteract the northern heating and so the rise in greenhouse gases is still forecast to produce a net warming around the North Atlantic. In experiments where the atmospheric greenhouse concentration is stabilised at twice its present-day value, the North Atlantic thermohaline circulation is projected to recover from the initial weakening within a few centuries. However, the models also indicate that a decrease of the thermohaline circulation reduces its resilience to climatic fluctuations, that is, a weakened circulation appears to be less stable and a shut-down could be more likely. But in spite of this, none of the current projections with the latest climate models predict a complete shut-down of the thermohaline circulation up to 2100.

Is there any evidence that global warming is affecting the thermohaline circulation at present? Certainly, deep waters at some locations in the northern North Atlantic now have a lower salt content than in previous decades. An attempt to monitor the northward flow in the Atlantic at 25°N has suggested the possibility that this flow might have declined, but some of this change might be no more than a seasonal difference. It therefore remains to be determined how widespread these changes have been, and the observations have not been made over sufficient years to compare the recent changes with natural climatic variability.

During the coming century the ecosystems in the northern hemisphere are likely to be subject to pressures caused by rising temperatures to which will be added those due to any weakening of the ocean conveyor belt. Although overall the current predictions are that the warming will be the strongest influence, the effects of the ocean circulation changes are still likely to be present. The previous chapter showed a climatic connection between the ocean circulation in the North Atlantic and distant biological communities, an association which suggests that the populations can be sensitive to what happens in the North Atlantic. If this is so, could a weakening of the ocean conveyor belt be accompanied by changes in the surrounding ecosystems quite distinct from the effects of global rises in temperature? Understanding of the climatic processes causing the connection between the position of the Gulf Stream and the abundance of plankton in the North Sea will therefore help in determining what ecological shifts might accompany a future weakening thermohaline circulation. Addressing this question involves exploring the processes by which weather affects biological populations.

Species are the atoms of which ecosystems are composed, and different species respond to their surroundings, both living and non-living, in individual ways. Identifying and classifying species is therefore basic to understanding ecosystems. As a consequence, the science of biological communities began with a bookkeeping exercise 300 years ago.

3

⌇

AT THE BEHEST
OF THE WEATHER

The seasons alter: hoary-headed frosts
Fall in the fresh laps of the crimson rose.

(William Shakespeare, *A Midsummer
Night's Dream*)

The North Wind doth blow and we shall have snow,
And what will poor robin do then, poor thing?
He'll sit in a barn and keep himself warm
and hide his head under his wing, poor thing.

(Traditional nursery rhyme)

When Europeans began to explore the rest of the world in the 18th century they brought back organisms from far-off lands, especially plants which could be easily grown in gardens and hothouses. These new samples, together with an increasing appreciation of natural flora and fauna, introduced the world of science to a bewildering diversity and richness. One man, Carl Linnaeus, a medical doctor with an interest in botany, decided to undertake the task of cataloguing all life as it was known at that time in the hope of providing some order to the variety. Linnaeus, a man not reticent about his own abilities, has been described as the first ecologist and the 'compleat naturalist'. His achievements have been summarised in the phrase: 'God created, Linnaeus arranged'.

Linnaeus began with the plants.[1] His great botanical work, *Species Plantarium*, published in 1753, after 20 years of labour, was a complete catalogue of all known plants. In it he employed the naming system, originally proposed by Aristotle, in which names were descriptive Latin phrases. This approach was quite cumbersome, so that the tomato had the botanical name of 'SOLANUM *caule inerme herbaceo, foliis pinnatis incisis, racemis simplicibus*' (SOLANUM with a smooth herbaceous stem, incised pinnate leaves and a simple inflorescence). In Latin, SOLANUM is defined as solace, referring to the narcotic properties of some related species. For convenience, in the margin of the text, Linnaeus adopted a system of abbreviated names which were simpler to remember: in this, the tomato now became *Solanum lycopersicum*. The name *lycopersicum* means 'wolf peach', and refers to the formerly supposed aphrodisiac powers of the fruit. These marginal jottings are now Linnaeus' main claim to fame: a bookkeeping scheme for filing species. Despite an initial cold reception from fellow scholars (one of whom in 1759 complained that dogs should not go with foxes and wolves but with horses, as both are found in the farmyard), the simplicity of the system recommended it to botanists and gardeners alike, and by the time Linnaeus died in 1778 his nomenclature was firmly established. All species are now named according to the scheme, and scientists even manage to slip some humour into the names: for example, *Ba humbugi* for a snail, and *Agra vation* and *Agra phobia* for a pair of beetles.[2]

Linnaeus' simple, concise method of naming organisms by genus and species revolutionised biology. These scientific names provide reference points in biological space to which researchers may compare future discoveries and information. By linking a name with a single 'type specimen', which can be examined again and again, biology became a repeatable science: the same species could be identified and recorded at different times and places. It is individual species that are responding to the vagaries of the atmosphere, even though several may show a similar response and the community as a whole may be affected. Knowing how different species live together is vital for understanding how they adjust to the weather.

The importance of individual species is exemplified by two species of copepods that are very common on the silks of the continuous plankton recorder. *Calanus finmarchicus* and *Calanus helgolandicus* are physically so similar that they were only recognised as separate species early on in the 20th century, and can be told apart solely by the structure of their fifth legs. Yet their lifestyles are very different. While *Calanus helgolandicus* struggles to continue breeding in the surface waters throughout the winter, its cousin, *Calanus finmarchicus*, lies dormant in cold water at a depth of several hundred metres, waiting to continue its life near the surface in the spring. In the North Atlantic, *Calanus helgolandicus* is a more southerly species than *Calanus finmarchicus*. These two animals are therefore dependent on different aspects of the weather and only *Calanus helgolandicus* is among the copepods tracking the Gulf Stream.

Once species could be listed, the question of why there are so many arose. This question was asked by the Yale ecologist G. Evelyn Hutchinson in a famous essay written in 1961. Although working in the USA, Hutchinson has been described as an ultra-eccentric, ultra-English don. He once sent a juvenile bushbaby hopping down the table in a faculty meeting to remind his somewhat annoyed molecularly biased colleagues about the importance of studying creatures. Forty years ago, Hutchinson was also publicising the notion that industrial carbon dioxide might lead to global warming and wondering if it mattered that when cows break wind, they emit methane. Plankton in the sea is very rich in species, so rich that Hutchinson called it the 'paradox of the plankton'. He argued that the ocean looks like a fairly homogeneous place being basically well-mixed water. Yet in any given region there may be thousands of different species of single-cell plants, all dependent on the same light and food supplies.

This seems to conflict with a general ecological principle.[3] Based on numerous experiments looking at how fruit flies, flour beetles, or mice respond to crowding, equations can be set up to predict what will happen if two different kinds of animal have to compete for the same kind of food in the same container. Intuition suggests that the two populations should achieve some kind of balance with more of one and less of the other. The mathematics, however, does not predict this common-sense outcome at all: there is total annihilation for one of the competitors and total victory for the other.

This mathematical prediction is in agreement with many laboratory experiments, the first of which were done at Moscow University during the grim days of the Stalin era. Georgii Frantsevich Gause set up numerous contests between pairs of different species of the single-celled animal *Paramecium* in glass tubes. Every time the outcome was the same: both populations would do well early on when there was plenty of room for all, but, as soon as they began to crowd, one species would start to decline and then disappear. In any repetition of an experiment it was always the same species that won. There have now been many other experiments like those of Gause, using many different kinds of animal and plants. Two species can only co-exist if somehow they are able to avoid one another. The consistent annihilation of one species is now called the competitive exclusion principle or Gause's law.

Yet, the seemingly uniform waters of the ocean seem to allow many species of plankton to exist side-by-side. This problem of the plankton is but one example of a common theme. Co-existence happens in tropical forests, where there may be a hundred kinds of tree packed together, and in an old pasture where a dozen kinds of grass and other plants will live mixed up together using the same water, relying on a common reservoir of nutrients, enduring the same seasons and experiencing the same accidents, and it happens in all the diverse arrays of animals. Why is this?

In the case of the plankton, one possible explanation is that the waters of the ocean are not as thoroughly mixed up as might be thought. The last chapter has already shown how strong ocean currents cause variations by transporting water from a very different geographical region, for example by moving warmer water into a cold water region. In addition, eddies are widespread and normally have physical and chemical constitutions that differ from the surrounding water-mass. Different patches of water could allow species to be separated and so not compete. But Hutchinson realised that this cannot be the complete explanation. His major field of expertise was limnology, the study of lakes (his four-volume *Treatise on Limnology* is a classic set of textbooks), and so he knew that even relatively small lakes have many species of phytoplankton drifting together in their open waters. Only the very largest lakes in the world, like the North American Great Lakes, contain currents or eddies which are anything like those in the ocean. A more convincing explanation is that species are able to exist side-by-side because each occupies a distinct *niche*.

An organism's ecological niche is how it makes its living: the resources it uses or prefers and the predators it seeks to avoid. Thus, a red fox, in the wild, frequently lives in forest edges and meadows, and is active at night, feeding on any small mammals, amphibians, insects and fruit it can find. At the same time, it may be bitten by blood sucking insects and is host to many diseases. All of this constitutes the fox's niche, a niche that is filled by the coyote in North America. By having niches that are not identical, species can avoid the competition that can lead to extinction. This is the main explanation for how organisms can appear to subsist almost on top of one another: if they are not kept apart, they occur at slightly different times or have individual food requirements, or else there is some other difference in their lifestyle.

One site, whose populations have been extensively studied over many years and so can illustrate this coexistence of species is Lake Windermere. Although the largest freshwater lake in the English Lake District, it is still only of moderate size. When I moved to the area in the 1970s the local Tourist Board called the Lake District 'The Roots of Heaven'. Certainly, when looking at Windermere surrounded by distant towering, snow-capped mountains in winter, the description seemed apt. The Lake is 17 km long, 1 km wide with an average depth of 20 m. Its deepest point is 65 m from the surface. Windermere is almost two lakes, consisting of a north basin and a south basin which are separated by an area of relative shallows (average depth 10 m) in the centre of the lake, where most of the islands are to be found. In each basin, the open water does not vary appreciably across the lake and the only currents within it are transient flows driven by the changing winds.

Windermere has contributed to the riddle of the oceanic plankton connection. When many of the observations in the opening chapter were being published in the early 1990s, it was already apparent that what was being seen was a response to weather patterns that were linked across the North Atlantic. The same association should therefore show up in data obtained on land. At that time I remembered that Glen George at the UK's Freshwater Biological Association had published observations of how the abundance of zooplankton had fluctuated in Windermere over the previous few decades.[4] Glen and I had once shared an office, and also a very frightening experience in a small boat on Windermere during a December gale. Glen's data showed that the zooplankton in Windermere have been connected to conditions in the North Atlantic by the same linkage as their cousins in the surrounding seas: as the Gulf Stream has shifted, so the zooplankton population has also risen and fallen.[5] The similarity between these two graphs (Fig. 3.1) is sufficiently strong that something like it would occur only about once in a hundred sets of purely random data.

Windermere is not unique in this: Glen has shown that the same connection has been happening in its neighbour, Esthwaite Water.[6] These observations are important in providing clear evidence that the pan-Atlantic connection operates through the atmosphere. There is no direct coupling between Windermere and the Ocean, so the weather patterns driving the plankton changes in the lake must have been coupled to those associated with the varying Gulf Stream position.

There is, however, an important difference from the linkage seen in the North Sea—the sign of the relationship is reversed. While in the North

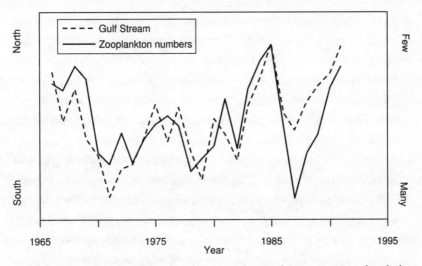

FIG 3.1 Summer zooplankton abundance in Lake Windermere compared with the latitude of the Gulf Stream. The zooplankton graph has been inverted (the graphs are in standardised units).

Sea northward positions of the Gulf Stream have been associated with *greater* abundance of zooplankton, in Windermere the association is with *smaller* populations. At the root of this difference are the processes responsible for the coexistence of phytoplankton species: different species can occur in different seasons, or even in different years. Conditions in the lake are constantly changing with time so that first one species blooms and then another. The tiny plants are short-lived and therefore can take turns occupying the water surface throughout the year. They might compete but never for long enough for a species to be eliminated. The zooplankton in Windermere depend on their particular food species being around at the right time.

Fluctuations in the populations of phytoplankton, and hence the zooplankton, are driven by the annual cycle of light and temperature in the lake in a way common to both lakes and seas. Indeed, observations from Windermere have provided much of the understanding of how these systems function. Many of these results came out of studies of the lake's fishery. This prolonged observational programme also illustrates some of the difficulties involved in keeping such a project going, and the value of the data sets that can be assembled.

Commercial exploitation of the fish populations in the lake can be traced back to the early medieval period when Furness Abbey owned much of the land. Perhaps the fishery helped ensure that the monks did not lack for fish on a Friday. From the 16th century until the end of the 19th century Windermere supported a commercial fishery with a steady yield each year of several tons of arctic char, trout, pike, perch and eels. The char were particularly valuable and were transported as far south as London, usually in the form of potted char. Although most fisheries faded out, perch fishing continued and in 1899 perch fishing was described as the main industry in the lakeside town of Bowness (a town that is now given over to tourists). Thereafter commercial fishing for 'coarse' fishes steadily became uneconomic. During the 1930s perch were abundant but mostly small. In an attempt to increase their size and provide a contribution to food production in the 1939–45 war, perch were trapped, canned in tomato sauce and sold under the name of 'perchines'. Reducing the numbers of fish in this way, together with the removal of some of the pike that preyed on them, gradually allowed the perch to grow faster and larger.[7]

This experiment on the ecosystem was carried out by the Freshwater Biological Association (FBA), a world-renowned research laboratory based on the Lake which had been set up in 1929. For the first few years the seven members of the scientific staff were housed at Wray Castle, a mock-Norman folly built in 1845 by a wealthy doctor from Liverpool. Subsequently, the Association took over a former Victorian hotel, the Ferry House, at the western end of the ferry crossing of the lake. Apart from its splendid views down the lake, this location was very convenient for members of staff catching the ferry at the end of the working day. In recent years, the main part of the laboratory became firstly the Institute of Freshwater Ecology and then part of the UK Centre for Ecology and Hydrology, before being moved away from the Ferry House.

Subsequently, it has shrunk because of funding cuts, and now almost all that remains is the core of the old FBA still located adjacent to the old hotel. Such can be the tribulations of any research laboratory whose life spans many decades.

As the perch experiment progressed, FBA staff monitored the ecosystem of Windermere in detail. They recorded the age and growth of individual perch and pike by measuring the annual rings that form in certain bones, much like reading the rings in a tree trunk. They also tagged and recaptured pike, providing estimates of population size and mortality. In so doing, a unique data set was generated that can now be exploited for other purposes by scientists elsewhere in the world. Thrond Haugen and colleagues from the University of Oslo used these detailed observations to examine how pike migrated about the Lake. They used these data to show that every year the pike in Windermere moved back and forth between the two basins away from the habitat with higher mortality to the one where they are more likely to survive. Although the south basin is more productive, pike survival in the south decreases more quickly with rising numbers than it does in the north, and so at some times of the year the pike preferred to be in the south basin and at other times in the north. In a separate study at the same university, the pike data have been used by Eric Edeline's group to demonstrate that, because large pike were selectively removed to help the perch population, the pike have gradually evolved during the experiment in favour of fish which grow more slowly and have a lower average size. This shift seems to be a common result of commercial fisheries.[8]

These two studies illustrate again how a research programme that continues for many years develops scientific potential well beyond its original purpose. The experimental programme extended further than the fish populations. Perch feed on insects and on fish such as sticklebacks and minnows, but the youngest perch live on zooplankton. This can be a critical time in the fish's life, being a period when they have limited food reserves, and so the programme included a twice-monthly sampling of the zooplankton, which was initiated in 1941 and has continued to the present day. In the early 1970s, I was fortunate enough to have an office at the FBA with spectacular views down the lake. I regularly saw the laboratory's blue and white motor launch *Velia* leaving to carry out this (and other) sampling. In those days the head boatman was a tall, quite silent man who was followed everywhere by his Jack Russell terrier that seemed to consider it its duty to make up for its master's lack of words by barking incessantly.

The preserved zooplankton samples from these expeditions have revealed that reducing fish stocks is not the only way that people have affected the lake's ecosystem. Over the decades up to 1991 there was a steady increase in the nutrients, in particular nitrate and phosphate, flowing into the lake, the result of agricultural runoff and the steady increase in tourism. This rise fed more intense blooms of phytoplankton during the summer, and these in turn could support larger populations of zooplankton. A steady increase in the numbers of zooplankton from year to year is observed in the zooplankton samples. The fluctuations shown in Fig. 3.1 were superimposed on top of this rising trend. The data shown in the graph stop in 1991. During that year, tertiary sewage treatment works were introduced on Windermere, which stopped the increasing nutrient inputs, thereby restoring the lake's health and gradually bringing an end to the rise in the zooplankton.

These observations of the zooplankton, and parallel observations of the single-celled plants on which they feed, reveal how the community present at any time is determined by the weather patterns that the lake has experienced. Many aspects of the chain of events are the same in most other lakes, and also in the seas and the oceans. These processes are why populations differ from one year to the next. The annual growth of the phytoplankton, which form the base of the food-chain in the open sea, is very dependent on how deep light penetrates into the water. If the plant cells are mixed downwards by the stirring in the lake, and so spend too long in the dark, they do not grow. Warming of the surface waters causes the density to vary down the water column and this gradient of density inhibits downwards transport.

In the spring, as the sun's power becomes greater, the lake or sea begins to warm up. However, because the heat input is to the surface water, only this is heated and, as the warm water expands, it floats on top of the colder water. Stirring action by the winds tends to spread this surface water into a thicker layer. By early summer the water column becomes thermally stratified and consists of an upper warm layer of water (called the *epilimnion* in lakes) floating on top of a deeper unheated layer (called the *hypolimnion* in lakes) with relatively little mixing between them. The transition between the warm and cold parts is called the thermocline. In Windermere this is normally located some five to ten metres below the surface during the summer, but in the ocean it is at 20–30 m or even deeper. During a good summer with frequent sunshine and little wind-stirring of the lake, the thermocline will be nearer the surface, whereas in a poor summer (cloudy and windy) the thermocline will be deeper. As autumn progresses, heat is lost from the surface, and the water column cools down. Eventually (commonly about November in Windermere), autumn gales have sufficient energy to remix the water column, and the temperature is once again the same at all depths throughout the winter months. This mixing extends to the bottom of Windermere, but in the ocean, which is much deeper, the mixing only gets down to a few hundred metres. This then sets up the conditions for the cycle to be repeated from the next spring.

Lakes like Windermere exist in many other countries. In London's National Gallery there is a striking blue, white, and silver painting by Alseli Gallen-Kallela showing the still waters of a lake with a central island surrounded by distant hills. The water reflects the cloud patterns and trees on the island, and is criss-crossed by slicks. The scene could easily be a view of Windermere but the title of the painting is 'Lake Ketiele' and the location is Finland. Any lake of sufficient size and depth will respond to the annual cycle of heating and cooling in the same way as Windermere. Indeed, most of the seas and oceans show a similar cycle of thermal stratification, although this may be modified by the mixing of the tides and the differences of density associated with varying salt content. Thermal stratification is one of the main reasons for the survival of a fish population in Windermere that is a relict of glacial times. During the long hot summer of 1976 surface temperatures in the lake reached 24°C but, even in this extreme year, the deepest water remained at less than 8°C, and this ensured that there was ample water for the Arctic char—whose eggs cannot develop at temperatures above 7°C—to survive in the lake. However, this fish will not survive in the lake if future warming means that the deep waters do not cool sufficiently in the winter.

Winds cause mixing in the deepest waters of lakes like Windermere, even during the summer. The prevailing winds over Windermere tend to funnel up the valley from the south, blowing the water in a northerly direction. During the summer, when the lake is thermally stratified, a strong south to westerly wind will drive warm surface water to the northern end of the lake. This piling up of surface water forces down the thermocline and drives deeper water in the hypolimnion southwards, causing the thermocline to tilt becoming deeper in the north and shallower in the south (the reverse happens in northerly winds). Once the wind stops blowing, the thermocline tries to regain its original position but in doing so it overshoots by several metres and oscillates back and forth every 12 to 15 hours until the rocking motion is damped out by friction or interfered with by the wind. This great wave propagates down the lake with scarcely a ripple at the surface. These movements within the lake are called an *internal seiche* (pronounced 'saysh') and occur to some extent in all lakes which stratify into two layers in the summer. In large lakes the Earth's rotation causes the wave to progress around the lake with a twisting motion. For organisms living freely in the water, the seiche is not too much of a problem because they can move with the water, but where there are fixed abstraction points, such as intake pipes, water temperature and quality can change markedly over a short period of time. Internal seiches are responsible for most of the movement and stirring in the deep waters of the lake. Waves like these also occur in the sea but are not as crucial to mixing as those in lakes.

In the sea, internal waves are responsible for *dead water*, a strange phenomenon occurring when a layer of fresh or brackish water rests on top of denser water without the two layers mixing. A ship in such conditions may be hard to manoeuvre or can slow down almost to a standstill. The phenomenon was first described by Fridtjof Nansen when he described how, near the Taymyr Peninsula, the *Fram* appeared to be held back as if by some mysterious force which sometimes stopped it answering the helm. Once again, an explanation was provided by Nansen's colleague, Walfrid Ekman. In dead water much of the energy of the ship's propeller is squandered in generating internal waves between the two layers of water, which leaves the ship capable of travelling only at a fraction of its normal speed.

This annual cycle of warming and cooling is how weather sets up the physical stage on which the living players in the lake or the sea act out their lives. Algae are the primary producers of the planktonic ecosystem, that is, like all plants, they have the capacity to convert inorganic material to biological matter using the energy of sunlight. Normally occurring as single cells or small colonies, they have a bewildering variety of forms in both fresh and marine waters. Thus, in Windermere, well over a hundred species have been recorded, a total which is almost certainly an underestimate of the true number. The different species of algae tend to spring up during the year in an orderly sequence determined by the seasonal changes in weather and the warming of the lake's waters. In the seas and oceans there is also a marked seasonality in the timing and appearance of different algal species, and these changes are broadly similar to those in lakes. Sunlight, heating, and wind stirring have similar effects on marine ecosystems to those in freshwaters. Therefore, the seasonal succession of phytoplankton in Windermere serves to illustrate the kinds of changes occurring in many other aquatic systems.

The first algae to bloom in the year are the diatoms, characterised by their delicate outer shell of silica. This is a form of glass and must appear as such to the animals that eat them. These algae appear first because they grow well at low temperatures and tolerate the low light levels that occur early in the year, particularly when the lake is fully mixed. In Windermere the spring bloom of diatoms is usually, but not always, dominated by *Asterionella formosa*, a colonial species which forms star-shaped clusters of cells. This diatom has links with Dickensian London. It was discovered and named 'the little star of beautiful form' by Dr Arthur Hill Hassall in the 19th century when he was investigating outbreaks of water-borne cholera and typhoid in water samples from the city. But Hassall's interest in the causes of sickness extended much wider than this.[9]

During the 1850s, Hassall examined some 2500 samples obtained from food outlets all over London. This was an age of unbridled capitalism, combined with widespread ignorance about the effects of chemical substances. Apart from alum and disgusting parasites in bread and sugar, he found chromates of lead and bisulphate of mercury used as colouring in children's sweets along with other compounds of copper, lead, and mercury, and strychnine and iron in beer. His investigations led to the 1875 Food and Drug Adulteration Act. Hassall subsequently moved on to work on another illness associated with Dickensian times, consumption. In 1866 he set up the Royal National Hospital for Consumption and Disease of the Chest on the UK's Isle of Wight; it was a pioneer in the field of sanatorium care for 80 years. Little now remains of the hospital, and its grounds have since become one of the gems of the island, the Ventnor Botanic Garden. The extensive paths are bordered by exotic plants from around the world and eventually lead down the cliffs to Steephill Cove where the sea breaks over the rocks amid small fishing boats. Even though engaged in all these other activities, Hassall still found time to write a definitive guide to British freshwater algae.

Diatoms dominate the oceans and are responsible for about a fifth of the primary productivity (plant growth) on Earth. Their growth in the spring is so prolific that they eventually consume all of the silica dissolved in the near-surface water. Thermal stratification is developing at this time and the density differences prevent the silica being replenished by mixing with the nutrient-rich waters below. Silica depletion causes the diatom population to collapse, die and sink to the lake bed. The loss of cells leads to a period of increased water clarity early in the summer, but this is soon followed by the growth of other species of algae. The summer phytoplankton community in Windermere is usually composed of *green algae*, many of which are surrounded by a mucilaginous sheath (a layer of slime). The term 'green algae', which is still used for freshwater phyto-plankton, is a remnant from when, in the early days of the science, algae were grouped according to colour. Nowadays, it is not even the case that all the algae in this grouping are green. The exact function of the sheath is unknown but, as the oils in it are buoyant, it may help to keep the cells afloat. This is important because these algae grow well at high light inten-sities and therefore need to remain in the upper, sunlit parts of the water column (the *euphotic* zone). Green algae are the food upon which some of the zooplankton depend.

In late summer the community of predominantly green algae begins to run out of other nutrients, mainly phosphorus and nitrogen, and as light intensity at the surface decreases with the onset of autumn, are replaced by what used to be known as *blue-green algae* but are now recognised as photosynthesizing bacteria and are now called 'cyanobacteria'. These 'pond-scum' organisms can thrive in warm water and grow at lower light levels. The name 'blue-greens' arose from the colour of the first such spe-cies to be recognised. At this time of the year, as nutrients are depleted at the surface, nutrient recycling in which the contents of dead cells are broken down by bacteria and returned to the water becomes a vital proc-ess. Some of these cyanobacteria have the additional ability of supple-menting the depleted nitrate in the water with nitrogen from the air. This is not the only use of gases by cyanobacteria and algae, as others contain gas 'bubbles' (vacuoles) which help to prevent them from sinking.

A similar annual cycle of change to that described for Windermere occurs in most of the world's seas and oceans. Although green and blue-green algae are terms that have been applied to freshwater phy-toplankton, there are equivalent marine populations. However, in all water bodies the sequence of phytoplankton through the year can some-times be more complicated than the simple progression just outlined if changing weather patterns alter the conditions for growth in the lake and different species of algae respond accordingly. The sequence may also be considerably different in lakes that are too shallow to become thermally stratified, or in seas that are prevented from stratifying by strong tides or winds.

One phytoplankton group that is different in the marine environment is the coccolithophores, and one species in particular, *Emiliania huxleyi*. These bloom late in the summer. Thus, for example, in the Gulf of Maine there is a diatom bloom in April, followed by dinoflagellates whose whip-like flagella are able to propel them back and forth between the surface, where they soak up sunlight, and a depth of 10–20 m, where they can spend the night absorbing nutrients. Then, the last phytoplankton to bloom are the coccolithophores, each of whose little cells are armoured in an ornate shield of chalk. Only electron microscopes can reveal just how intricate and beautiful is the structure of this shell. It is made up of tiny delicate plates called coccoliths, often having the appearance of wheels with spokes. In late June, parts of the Gulf of Maine turn milky white as the phytoplankton shed these coccoliths in vast numbers when their bloom is coming to its end. Such intense blooms of *Emiliania huxleyi* occur over large areas of the North Atlantic Ocean, the pale waters being seen clearly from space and showing up distinctly on satellite images of the ocean. The species also occurs to a lesser extent in many other parts of the world's oceans and coastal seas. Accumulated coccoliths from ancient blooms have formed massive deposits of chalk, such as the White Cliffs of Dover and those under the Champagne region of France where artificial caves have been dug out of the chalk for storage of wine.

Phytoplankton are grazed by zooplankton which have more complicated lifestyles and are in turn grazed upon by larger animals. A common pattern of zooplankton behaviour is for the animals to be at the surface by night but to secrete themselves at great depths during the day. This migration may be to avoid predators, but a number of other explanations have been suggested. The numbers of zooplankton vary greatly throughout the year, sparse in winter but swarming much more densely in summer. This annual cycle is the reason that the connection with the Gulf Stream in Windermere differs from that seen in the North Sea. In the marine system the zooplankton is made up of a number of species, commonly copepods, appearing at slightly different times of the year and eating different algal foods. If the summer is a good one, the phytoplankton crops are strong and the zooplankton are able to take advantage of them and reproduce abundantly. Here, a good year for phytoplankton means lots of copepods.

In Windermere, the zooplankton is made up of small invertebrate animals of various types but at the peak of the zooplankton during May and June the population is dominated by a single species, *Daphnia hyalina-galeata*. *Daphnia* are commonly referred to as 'water-fleas' even though they are neither parasitic nor fleas. Like many planktonic animals, these water-fleas have adopted filter-feeding lifestyles and they strain small food particles, such as algae, from the water. Indeed, during the summer when they dominate the community, this grazing effect can exert a significant control over the numbers of algae in the lake. The growth of *Daphnia hyalina-galeata* is physiologically timed so that the expanding population can feed on the bloom of green algae, and this has the consequence that the size of its population is dependent on how the algae are affected by the weather.

When the spring months are dominated by calm, dry periods there is more intense stratification in the lake and this bloom is past its peak before the *Daphnia* can make use of it. Although the *Daphnia* population also takes off earlier in the warmer conditions, the phytoplankton 'out accelerate' it and, once the animals have missed the boat, they never really recover during the weeks that follow. In this way, the annual succession leads to a poor zooplankton population occurring during a good summer. Zooplankton abundance is higher in years when the weather is cool during the spring and early summer and lower in years when the months of May and June are warm.[4] This mismatch of timing is the source of the difference between the animals of Windermere and those in the North Sea.

Phytoplankton co-existence in Windermere therefore determines how fluctuating weather affects population sizes. Zooplankton in the lake are biologically programmed to feed on algae occupying a specific niche in the annual cycle. Growth of the green algae is critically dependent on the collapse of the diatom bloom. The *Daphnia* do increase earlier when the temperature increases but they are unable to follow the algae. How different species in lakes and seas survive side-by-side in a place which seems to have nowhere to hide can therefore be important for what happens when weather patterns change.

While the homogeneity of the environment does not at first sight appear to allow large number of species to co-exist in aquatic systems, there is no such uniformity on land. Although regions may sometimes seem to be uniform, they are broken up by rivers, boulders, mountains, and rock and soil types in ways that allow a wide variety of plants and animals to thrive. Even so, there are still smaller areas such as grassland or woods uniform enough that species could be in direct competition for the same resources.

Terrestrial species interact and are affected by their environment in very different ways from those living in water. For a start, plants on land are commonly dependent on rainfall to provide them with water. Land plants and the animals around them are subject to stresses, such as winter frosts or scorching temperatures, not felt by aquatic organisms. In addition, unlike algae which are susceptible to the weather on almost a day-by-day basis, terrestrial plants have much longer lifecycles and slower responses. Many of these differences between terrestrial and surrounding ecosystems are illustrated by a set of observations of wild plants that has been assembled in the English Cotswolds.

The Cotswolds are a range of English hills stretching nearly 100 km from the city of Bath into Gloucestershire. The country is mostly open down-land, but here and there small patches of woodland can be found, crowning a hill or covering a slope. In among the villages are old houses, solidly built of hard, mellow limestone with a yellowish-grey hue. The underlying rocks from which this mustard stone was extracted are from the Middle Jurassic, a geological period named after the Jura Mountains in Switzerland. These 170-million-year-old rocks were originally thin sediments deposited in shallow seas and estuaries, in basins that were subjected to frequent earthquakes and faults, uplifts, and down-thrusts. Fossilised animals and plants from these events have played an important role in the development of the science of geology.

In the 1790s William Smith, a young man from simple yeoman stock with a passion for rocks and fossils, was employed surveying the route of a canal south-west of Bath. His observations of the strata of the rocks along the route led him to realise that, by examining the fossils present in the different layers, it would be possible to map the geology of the area. Thus began a 20-year obsession that would produce the first map of the geology of Britain (or anywhere else). His vision cost him dear for during the course of his work his wife went insane, his completed map was stolen by jealous colleagues who eventually ruined him, and he was imprisoned for debt. It was some years before he got the credit to which he was entitled for producing a map that fundamentally changed the way we view the world.[10]

On the way to taking up his canal-surveying post, William Smith travelled along the roads the Romans had built: striking south-west along the remains of Akeman Street before turning on a more southerly route via the Fosse Way towards Bath. Fosse Way is particularly remarkable, for over its direct course of 180 miles from Lincoln to Somerset, no part is more than six miles away from a straight line joining the extreme points. Fosse Way and Akeman Street enter the north-east gate of the Roman city of Cirencester together.

Not far from Cirencester, Akeman Street passes close to the village of Bibury and here on one of its verges Arthur Willis and colleagues from Sheffield University have assembled a unique set of observations of the growth of wild plants over several decades. Bibury is an idyllic spot for a scientific experiment. The famous Victorian designer William Morris once described it as 'the most beautiful village in England'. It is surrounded by hills, and straddles the gently flowing River Coln, beside which is a much-visited row of traditional cottages. This line of weavers' cottages, overlooking the mill stream, is called Arlington Row, and is one of the most photographed places in the region. The village grew rich on wool and milling, so rich that the 'living' of the vicar was one of the best appointments in the Cotswolds. Naturally the vicars tended to stay, with the result that there were only 12 in the 400 years from the mid 15th century, a period when the country was ruled by almost twice as many monarchs.

Arthur Willis was a scientist with a life-long fascination for plants, which developed during his childhood on a farm. There he learned the skills of milking, haymaking, and driving a horse and cart. He also had a nearly fatal encounter with nature when his pram tipped over and he landed in a slurry heap. The monitoring programme at Bibury began by accident when Willis' group were carrying out a series of tests of different weedkillers during the 1960s and 1970s. Along with each treatment they recorded in detail the plant composition of other patches of ground to which nothing was added, so that any effects of the herbicides could be properly assessed. Their continuous annual monitoring of these untreated plots of the verge as controls gradually developed into a long-running series of observations of how all the wild plants in the verge varied over the subsequent four decades.

This roadside community was not subjected to ploughing or herbicides like those in surrounding fields. Realising the potential significance of this long-running set of data, Willis carried on making the observations, continuing well into his retirement. The simple procedure used since the beginning of the programme is to identify and count all the plants in several $1\,m^2$ areas of the verge at a fixed time of the year in July, not long before the verges were given their annual cut at the end of the summer. Willis' work may well represent the world's longest running dataset for land vegetation continuously involving a single individual using the same methodology.[11]

The Bibury long-term data set contains information on fluctuations in the abundance of over 100 grasses and wild plants over 40 years, and so represents a unique record of the dynamics of a plant community. Species are jostling for survival, shoulder to shoulder, each plant existing because it has a unique niche defined by whether it prefers sun or shade, how much or little rain it likes, its method of pollination and seed dispersal, the time of year it appears, and many other aspects of its life. How well the plant does under the weather in any particular year is decided by the nature of its niche.

Glen George learned of this suite of data when he was involved with Sheffield University as one of the partners in a collaborative research programme called TIGER. These plants have been subject to weather patterns connected to those in Windermere and the North Sea, and Glen suggested seeking signs of the pan-Atlantic linkage with the Gulf Stream in the Bibury observations. The connection underlies many of the fluctuations along the roadside as it does in the communities of the seas and lakes. Figure 3.2 shows how one of these roadside plant populations, field scabious, has been tracking the position of the Gulf Stream. This connection is seen to varying degrees with many of the 40 or so commonest species and, by analysing these relationships statistically, Arthur Willis and his colleagues showed similarities like that shown in the figure occurred both more often and more strongly than would be expected to happen by chance.[12] There is, however, a significant difference from the aquatic systems: not all the responses are in the same direction. When some species experience a good year, others find the conditions difficult.

A range of meteorological factors are responsible for the marked variations in how much different species grow in any individual year but, for the most abundant species, it seems that weather patterns in the spring tended to have more impact than those in any other season.[13] In particular, temperature, which acts as the limiting factor on the rate and timing of shoot growth during the spring, seems to be particularly important. Total vegetation production tended to increase with minimum spring temperature. In general, those species which cope best with environmental distress or disturbance tended to do better following warm dry springs and summers, whereas those favoured by more productive conditions

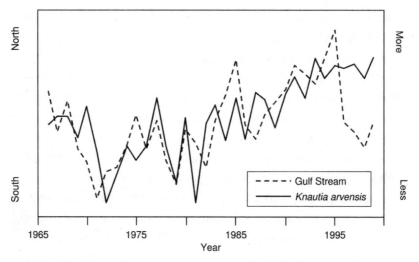

FIG 3.2 Abundance of one common species on the roadside verge at Bibury compared with the latitude of the Gulf Stream (the graphs are in standardised units).

were promoted following a wet growing season. All this implies that the connection with the distant ocean operates through several aspects of the weather.

Another major cause of the differing responses of individual species is competition. When two members of the group, Nigel Dunnett and Philip Grime, provided five of the common roadside-verge species with warming in the spring, they all responded with increased height and greater ground-cover if they were grown alone. However, this uniformity of response was not observed when the species were grown together. Competition between the species sustained the benefit of the warming in two of the species but negated the potential benefit in two others. These effects could still be seen in the following season. Such competition may even amplify the responses of vegetation to changes in the climate by favouring some plants at the expense of others.[14]

The competitive struggle between plants happens in all terrestrial ecosystems: it even occurs in the harsh environments of deserts. The tall, branched cacti that form a backdrop to many westerns are saguaros (*Carnegiea gigantea*). Widespread through southern Arizona, these very slow-growing plants typically take well over 100 years to reach the size seen on the cinema screen. In the course of its lifespan a healthy saguaro produces about 40 million viable seeds of which only a tiny fraction avoid being destroyed by harvester ants, birds, and rodents, and most of the tender seedlings that do germinate are also eaten in the first year. The majority of the seedlings that manage to survive are in sheltered places underneath paloverde trees and other plants. Groups of several saguaros may often be seen beneath, and rising through, a single paloverde tree or some other *nurse plant*. Growth in these early years is slow: it probably takes about ten years to reach a height of about 2 cm and 20–50 years to reach 1 m. But the competitive struggle has only been delayed. Gradually, as the saguaro grows, its shallow roots spread out and consume the water needed by the sheltering tree or shrub, which then dies and is succeeded by the mature cactus.[15]

The roadside verge at Bibury is an established community of plants with any gaps being continuously filled by seeds from plants growing in neighbouring hedgerows and fields, or else by new shoots sprouting from the roots of surrounding plants. Plants may also spring up from seeds that are not fresh arrivals but have instead been awaiting their opportunity within the soil, sometimes for a long time. Following the economic collapse after the Franco-Prussian war of 1870, the wheat fields of northern France were abandoned and stayed grazed and flowerless until 1914, remaining largely free of bushes and shrubs. In the days of wheat growing, the crop was full of weeds, and poppies were everywhere. When the shells of the First World War ploughed up the fields again and stopped the grazing, the poppies bloomed once more, after lying in the soil for nearly half a century. These fields of poppies have since come to symbolise the tragedy of war.[16]

A community like that at Bibury has been achieved by process of succession, much slower but otherwise analogous to that seen in lakes and seas. The succession operates by a competitive struggle of the kind described, with one group of plants elbowing out their neighbours. Any area of bare ground, such as freshly exposed rock, or an area cleared of vegetation by fire, flood, or human activity, does not remain empty of vegetation for long: an uncultivated field is quickly invaded by weeds, and sand dunes are colonised and stabilised by marram grass or similar specialised drought-resisting grasses. Such succession where vegetation has not been grown before is called primary succession (as opposed to secondary succession when an established ecological community is disturbed so that it undergoes alterations in its structure). These early pioneer species are those that are able to colonise open, exposed sites, free of competition from other species. The rapidity with which plants and animals can recolonise is illustrated by a particularly dramatic example.

When the volcano Krakatoa, lying between Java and Sumatra, blew up in 1883 after a long series of repeated eruptions at least half of its island disappeared beneath the ocean and the remainder was covered by ash and pumice. All its animal and plant life was obliterated. But the island was recolonised surprisingly quickly. Nine months after the eruption a single small spider had ballooned across from the mainland on a silk thread. By 1905, the first reticulated python had arrived, and after 25 years nearly 200 species of insects had been identified, and the forest was beginning to grow, although there were still no mammals. The black rat and several other plants arrived by boat with the scientists and tourists. One unanswered question has been the growth of fig trees. Fig plants can only be pollinated by female wasps of a particular species (one for each fig species). So how do fig trees colonise an island without the wasps they need, and how do the wasps colonise an island without the figs they need?[17] This question remains unanswered.

Pioneer species do not generally last for long, however. In an abandoned arable field, pioneer weed communities will be quickly replaced by grassland which, if left ungrazed for a few years, will be invaded by bushes and converted into scrub. The first weeds to arrive are annual plants that have the ability to spread tiny seeds far and wide on the off-chance that there will be a patch of bare earth in which they can grow for a season and scatter some more seeds. Next come perennials, herbs that drive resistant root systems into the ground and hold onto the fields year after year. These are followed by choking thorn-bushes and then the scrub trees of woodlands. The scrub community may last longer than the weed and grass plant communities, but eventually it will be colonised and overgrown by tree species and so changed to woodland.

Ecologists in North America and Europe in the early part of the 20th century used to think that plant succession in any region tended to end up with a particular vegetation type. In the case of the abandoned field this ultimate community is woodland. The final stage in succession, the climax vegetation had a fixed and predictable composition which was directly related to, and in balance with, its physical environment. It was supposed to be dependent principally on the weather regime of the area, so that the global distribution of vegetation was determined by the world's climate. Climate does exert a powerful influence on vegetation: tundra and tropical rainforest vegetation will not develop in each other's climates, and neither will develop in those of the Mediterranean. However, the detailed nature of climax vegetation is the result of a number of factors which also include some less obvious ones such as relief, soil and drainage, the nature of parent materials, and the influence of animal and human factors.

It is well known, especially to gardeners, that soils tend to reflect the plants growing on them. Soils determine the flora they support; for instance, only certain plants will grow on chalk. But plants also modify the soil and so some of the close link between vegetation and soils can be explained by the relationships that govern succession. The initial establishment of vegetation often begins the process of soil stabilisation. Pioneer communities are able, by root action, to bind soil particles together and thus to prevent the removal of soil by wind and rain. A developing cover of leafy vegetation also reduces the direct impact of rainfall and restricts wind speed at the ground surface, thus lessening soil erosion. The protection given to the soil by invading plants is crucial to continuing succession: without it successions would not proceed and plants would not remain long enough to impress their characteristics on the soil. Vegetation also alters the properties of soil by adding organic matter

75

which holds water and nutrients, such as nitrogen, phosphorus, and potassium. Plants therefore help to alter their local environment or habitat. In so doing, they modify the conditions of the site to such an extent that other species (which may have been unable to colonise because of lack of water or nutrients, or because of exposure) may now find the habitat suitable. Once these later species become established in a succession, they continue to modify their local environment and so encourage further succession.

This is the kind of struggle going on from year-to-year in the roadside verge at Bibury. Plants in the community are trying to replace other plants, but many are unable to do so because of the annual cropping of the verge, which prevents shrubs and trees gaining a foothold. How much progress any individual plant makes before this happens is determined by the weather. The long series of observations made by Willis' group has shown that from one year to the next the weather caused temporary adjustments to the balance of the verge with some plants growing at the expense of others.

Many of the species present are perennials (the field scabious in Fig. 3.2 is an example). These survive through the winter from one year to the next and, as a consequence, disturbances caused by the weather may carry over into the next year. The effects of one good or bad summer may be felt into the following year and beyond so that the proliferation of these plants is determined by the weather of several seasons. It is therefore not difficult to see how a connection with movements of the Gulf Stream might have arisen. Both the plants' growth and the latitude of the Gulf Stream are an accumulation of weather patterns over several years, and the weather systems concerned are part of the atmospheric circulation that spans the North Atlantic from the seaboard of the USA to Europe. The association between this living community and the ocean currents happens because the ecosystem and the Gulf Stream are responding to different parts of large-scale atmospheric patterns spanning several years.

However, there remains the question of how the connection appears in the plankton. Phytoplankton cells have generation times of days and are influenced by weather events on a daily timescale, and although zooplankton have longer generation times, typically weeks, they are still susceptible to what happens to the weather from week to week. Neither group of plankton has much recollection of weather patterns over years in the way that happens in the ocean circulation. Yet zooplankton numbers, which are determined by the weather over a few weeks or months, have been going up and down in line with the position of the Gulf Stream, which is the result of weather patterns over much longer than a year. How has this been happening?

A possible explanation is that the planktonic ecosystem is in some way sensitive to weather changes that originate over the ocean and are initiated by the oceanic conditions. What does the ocean do to the weather? Fundamental to how oceanic processes might lead to connections spanning the North Atlantic Ocean is the nature of the world's weather. A dynamic network of interconnections links weather patterns around the globe and these atmospheric forces are the source of climatic changes across the region. However, a complication in understanding all these processes is that the weather systems and ecosystems are non-linear, that is they do not necessarily react in a straightforward way to anything that occurs. This imposes strict limits to all forecasting. So, before embarking on an examination of atmospheric systems and their impacts, the next chapter considers the issue of Cleopatra's nose.

4

⌣⌣

THE SNAKE IN THE MATHEMATICAL GRASS

'Had Cleopatra's nose been shorter, the whole aspect of the world would have been altered.'

(*Pensées*, Blaise Pascal)

The Hudson's Bay Company is one of the oldest, still-active companies in the world, being already almost 200 years old when Canada was established in 1867. From its inception in 1670, the company controlled an area, designated Rupert's Land, which is one-third of present-day Canadian territory. Control over this enormous domain was granted by Royal Charter following the successful voyage of the *Nonsuch* to trade for beaver pelts with the Cree near James Bay. What began as a simple fur-trading enterprise evolved into a trading and exploration company that is today one of Canada's largest retailers. Less well known is the fact that the Hudson's Bay Company has also made a significant contribution to the science of ecology. This began in 1937 when D.A. MacLulich carried out an analysis of numbers of animal pelts shipped from the company to London between 1849 and 1904.[1] These pelts came primarily from around the James Bay area because

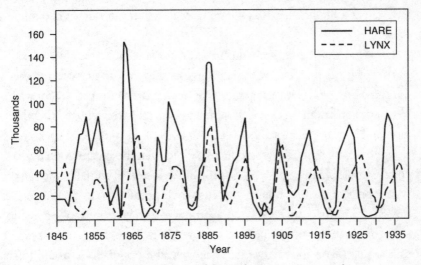

FIG 4.1 Trading figures of the Hudson's Bay Company.[2]

the pelts were not valuable enough to be shipped over the additional distance from western Canada. The numbers of pelts revealed that, throughout the 19th and early 20th centuries, the populations of two species, the snowshoe hare and the Canada lynx, have fluctuated cyclically about every 10 years. Populations of the hare can reach 8 hares per hectare and fall to one hare per 10 hectares—an 80-fold range of densities. The lynx numbers followed those of the hares but with a delay of a year or two.

Although the exact cause remains unclear, all the signs are that these oscillations are attributable to the biology of the animals themselves and not due to weather events: there is no evidence for the weather cycling in such a manner. The snowshoe hare lives in the northern forests from Alaska to the Atlantic feeding on whatever vegetation is available throughout the year. It has up to four litters a year and a lifespan in the wild of four or five years. The Canada lynx is a wild cat living in the same forest and preying almost exclusively on the hares. It is secretive and most active at night, so that it is rarely seen in the wild, and even trappers who have spent a lifetime in areas where lynx are common rarely encounter these predators.

The maximum hare population has been constant to within a factor of two over the nine cycles, which suggests that the fluctuations are the outcome of some relatively stable cycle. Some part of the ecosystem was putting a constraint on the peak numbers. MacLulich speculated that the hare and lynx populations were causing each other to cycle. When lynx were rare, hare numbers would increase and then the greater food availability would enable the lynx to reproduce faster. Eventually, the lynx would be eating enough hares to cause their population to decline, and this would in turn drastically reduce the lynx growth rate causing a fall in their numbers. With many of the predators gone, the hare population would begin to rise again. In this way, the populations would follow the repeating cycles that are observed. But subsequent investigators have pointed out that the system is more complex than this. As the hares approach peak densities they may well start to eat more than the plants can tolerate. If so, the food available will drop below what is required to keep the hare population alive through the winter. In addition, the hares are preyed upon by other predators such as arctic foxes, and are vulnerable to avian predators when outside the forest. In any event, it seems that the oscillations of these species are part of a tapestry woven from the complex interactions within the system. Over the nine cycles of the populations there seemed to be no changes in the weather patterns that could account for the oscillations in the system. Instead, the cycles occurred because of the subtle interactions between the hares and the lynx, and between the hares and their food supply. The cycles are almost certainly a manifestation of the non-linearity of these interactions. They are an illustration of non-linear processes creating an unexpected phenomenon. Processes in which the effect increases in proportion to its inputs are called linear. Thus, doubling the speed of a car doubles the distance it travels in a given time (traffic permitting). Linear relationships can be captured by a straight line on a graph and are generally not too difficult to handle. Their equations are the easiest to solve, with the result that these are the phenomena best understood and so are the ones that appear in textbooks. Linear systems also have another important property: they are modular. You can take them apart and put them back together again—the whole is equal to the sum of the parts.

But most processes in nature are not like this, they either do not change proportionately or even not always in the same direction. Often contaminants that are toxic to organisms can still actively promote growth at low concentrations. This phenomenon is called *hormesis*. As with all plants and animals, plankton in the seas or lakes grow slower when their food is in short supply, but their growth is restricted to a maximum rate once food is in excess. Further increases in food availability have no effect. Plankton are also tied in to the cycle of thermal stratification as the water column warms up, which depends on the interaction of sun, winds, and heat loss from the water surface. The long-distance connection in the opening chapter therefore associates the Gulf Stream, a current driven by complex physical forces, with systems that are rife with non-linear processes. The link is a light shining through a fog of complications. Before attempting an analysis of this connection, it is necessary to appreciate the effects of these kinds of processes.

Non-linear systems generally cannot be easily treated mathematically and cannot be subdivided into distinct parts. As a consequence, these non-linear terms tend to be the processes that scientists leave out when they try to get an elementary understanding. Non-linearity is present throughout the earth's physical and biological system, although its importance has not always been appreciated. While its effects are hard to calculate, it also creates rich kinds of behaviour that are not seen in linear systems.

One of the earliest examples of this was observed over 150 years ago when a young Scottish engineer named John Scott Russell was conducting experiments to determine the most efficient design for canal boats.[3] After testing boats on the Union Canal near Edinburgh, he decided it was the great bow waves that the boats made that was slowing them down.

Ten years later at the 1844 meeting of the British Association for the Advancement of Science in York, he described what happened next:

> I was observing the motion of a boat which was rapidly drawn along a narrow channel by a pair of horses, when the boat suddenly stopped—not so the mass of water in the channel which it had put in motion; it accumulated round the prow of the vessel in a state of violent agitation, then suddenly leaving it behind, rolled forward with great velocity, assuming the form of a large solitary elevation, a rounded, smooth and well-defined heap of water, which continued its course along the channel apparently without change of form or diminution of speed. I followed it on horseback, and overtook it still rolling on at a rate of some eight or nine miles an hour, preserving its original figure some thirty feet long and a foot to a foot and a half in height. Its height gradually diminished, and after a chase of one or two miles I lost it in the windings of the channel. Such, in the month of August 1834, was my first chance interview with that singular and beautiful phenomenon which I have called the Wave of Translation.[4]

Russell knew that this was no ordinary wave. Since the days of Newton, scientists had described exactly how waves travel and disperse. They knew that in the dynamics of fluids everything boils down to one set of equations, the Navier–Stokes equations. These relate a fluid's velocity, pressure, density, and viscosity in a precise way, but they are strongly non-linear. For most 'normal' waves, their height is sufficiently small and their currents slow enough that such complexities can be neglected. These waves gradually disperse and flatten out. However, if they run into shallow water they can steepen and topple over, as happens when waves break on a beach. Russell's wave of translation, having only a single hump is a *solitary wave*. It is so stable because these opposing tendencies of dispersing and steepening are continuously in balance, so cancelling each other out. The non-linear processes that lie at the heart of steepening and breaking are crucial to these waves.

Following his observation on the canal, Russell built a 30-foot tank in his back garden and began three years of experiments on the waves. He discovered that the solitary wave did all sorts of strange things. Firstly, the speed of the wave increases with its height, and its wavelength is determined by the depth of the water. But most unusually, they are incredibly stable: not only do they keep going for miles, but unlike normal waves they never merge. A small wave is merely overtaken by a large one, rather than the two combining. Russell surmised that the tides might behave just like these waves and hoped that this knowledge might help to improve coastal defences and tidal rivers. He also found that if a canal boat was pulled at just the right speed it could rise up on the solitary wave and surf comfortably along with very little effort.

But despite his insistence that his discovery was important, no one really wanted to know, and his work was almost totally forgotten for a century. Scientists of his day were just not prepared to accept what appeared to be a contradiction to the accepted theories of hydrodynamics. It was not until the mid 1960s, when scientists began to use computers to study non-linear processes, that Russell's early ideas began to be appreciated. He viewed the solitary wave as a self-sufficient dynamic entity displaying many properties of a particle, and many recent applications have built on this idea. Russell's wave, now called a *soliton*, has since been used to understand the complex dynamical behaviour of wave systems in plasmas, shock waves, tornados, elementary particles, and even the Great Red Spot of Jupiter. Today's fibre-optic communications use stable pulses of light identical to Russell's waves to carry masses of information over thousands of kilometres of fibres.

Unfortunately, Russell was a better scientist than a business man. He went on to organise the Royal Commission for the Great Exhibition of 1851 and to help design Britain's first armoured warship, the *Warrior*. But when he teamed up with Isambard Kingdom Brunel to build the colossal *Great Eastern*, they fell out badly over finances. Subsequently, Russell's business gradually declined and, despite his earlier successes, he eventually died a relatively poor man in 1882.

Freak waves are not confined to canals. In June 2008, a Japanese fishing boat, the *Suwa-Maru No. 58*, capsized in the Pacific in apparently moderate seas, killing 17 of its 20 crew. Investigators of the tragedy conjectured that it had been hit by sudden big waves. Over the years, there have been many reports of such freak waves, especially in the region of St Johns, South Africa where the Agulhas current interacts with other currents. By reconstructing the wind and sea conditions at the time of the event in a computer model, Hitoshi Tamura and his colleagues at the Japan Agency for Marine-Earth Science and Technology in Yokohama have been able to offer a plausible account of how such mysterious waves form. Non-linear coupling between the swell and wind-waves caused the low and high frequency components of ordinary ocean waves to interact, channelling their energy into a narrow frequency band, thereby creating waves with very large amplitudes.[5]

The *Suwa-Maru No. 58* is not the only ship to have been lost in such circumstances: anecdotal evidence from mariners' testimonies, and damage inflicted on ships, have long suggested that such freak waves occur. And it is not just the height of the wave that damages ships; vessels can suddenly drop into the trough before the wave. One of the oldest freak-wave reports was of the lighthouse on Eagle Island off the west coast of Ireland being struck by the sea on 11th March 1861, smashing 23 panes, washing some of the lamps down the stairs and damaging the reflectors beyond repair.[6] In order to damage the uppermost part of the lighthouse, the water would have had to surmount a seaside cliff measuring 40 m and a further 26 m of lighthouse structure. However, such waves remained almost the subject of legends until a rogue wave with a height of 25.6 m struck and damaged the Draupner oil platform in the North Sea on 1 January 1995 and was captured on recording equipment (Fig. 4.2).

Just as the oscillations of hares and lynx are a tapestry woven from the non-linear threads in their ecosystem, so Russell's solitary wave and these freak waves are others woven out of the dynamics of how fluids flow. These particular tapestries still each represent a relatively simple weaving of the threads. Much more intricate patterns can occur in other systems. In the oceans, the intricacies of the physical processes cause the shedding of eddies from ocean currents, and the Gulf Stream to abruptly break away from the seaboard of North America at Cape Hatteras. Non-linear processes therefore influence both ends of the

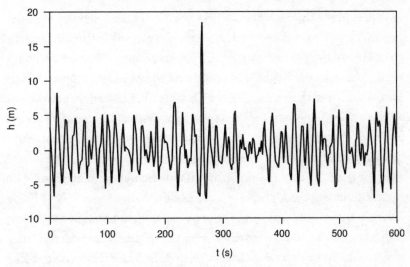

FIG 4.2 Wave-heights (m) at the Draupner oil platform in the North Sea on 1 January, 1995.

pan-Atlantic connection as well as permeating the atmospheric linkage between them. But the effects are not restricted to waves and oscillations; the most important effects attributable to non-linear processes are the severe limits they impose on predictability. Such limits arise when the way a system develops is extremely sensitive to how it starts out, and they restrict the forecasting of both weather and biological systems.[7] This kind of sensitivity is now understood to be widespread in nature. Although commonly a feature of relatively complex systems, one of the earliest descriptions of it happening was in a simple, small-scale experiment.

D'Arcy Wentworth Thompson was a British biologist and naturalist from the early years of the 20th century whose most famous work, *On Growth and Form*,[8] attempted to reduce the patterns observed in living organisms to a system of generating forces. He wrote, 'It may be that all the laws of energy, and all the properties of matter, and all the chemistry of all the colloids are as powerless to explain the body as they are impotent to comprehend the soul. For my part, I think not.' Because he played little part in the chemistry or genetics of cells, the 20th century's revolution in biology passed him by, but even so he was still extremely influential. Many of the greatest biologists have found themselves drawn to his book. The palaeontologist and celebrated writer on evolution Stephen Jay Gould wrote that before D'Arcy Thompson, 'Few had asked whether all the patterns might be reduced to a single system of generating forces, and few seemed to sense what significance such a proof of unity might possess for the science of organic form.'

In his book, Thompson points out the threads and columns made by ink drops falling through water look like the hanging tendrils of living jellyfish. If the similarity between sinuous threads made by the falling droplets and the living tendrils is more than just a coincidence, they may be caused by processes that are related. Thompson goes on to make a further observation: 'An extremely curious result…is to show how sensitive these…drops are to physical conditions. For using the same gelatine all the while, and merely varying the density of the fluid in the third decimal place, we obtain a whole range of configurations, from the ordinary hanging drop to the same with a ribbed pattern…'[7,8] The behaviour of the ink drops therefore suggests that the configuration of jellyfish tendrils could also be highly sensitive to changes in their liquid environment and their formative processes.

Whether or not jellyfish are strongly influenced by the tiniest details in their surroundings, it was discovered in the early days of computer modelling that such sensitivity is certainly a common feature of the atmosphere. In 1960, Edward Lorenz developed a simple meteorological model which ran on an electronic computer at Massachusetts Institute of Technology. Within his primitive computer he had reduced the atmospheric system to the barest skeleton, but even so the winds and temperatures in his printouts seemed to behave as recognisable weather patterns. One day in the winter of 1961, wanting to run the model for a longer period, Lorenz took a shortcut. Instead of repeating the earlier part of the run, he started in the middle, and, to provide the weather patterns at the time when he began, he typed the numbers straight from an earlier printout.

Lorenz then went away and left the computer calculating. When he came back and looked at the results he saw something unexpected. This new run should have duplicated the old run and gone on to extend it, but instead the weather patterns rapidly grew farther and farther from the original run, until after a few months all resemblance had disappeared.[7]

Lorenz quickly realised what had happened. Although six decimal places had been stored in the computer's memory, he had entered only three, and so his initial condition was slightly different from the condition in the middle of the run. He had assumed that the difference, one part in a thousand, was inconsequential. But in reality the tiny differences had grown to produce patterns unrelated to the original. He and others quickly grasped that this behaviour was a critical property of the global atmosphere. Tiny fluctuations in the atmosphere grow with time to have major ramifications. In 1972, Lorenz wrote a paper with the title: 'Predict-ability: Does the flap of a butterfly's wings in Brazil set off a tornado in Texas?' Following this, the tendency for seemingly inconsequential dis-turbances in the atmosphere to become amplified and give rise to larger, even catastrophic, effects has become widely known as the *butterfly effect*. It is an irrevocable result of the non-linearities present in the interactions of the atmospheric system and limits detailed weather predictions to no more than a week or so ahead.

The cosmologist Alan Guth has even light-heartedly argued that the weather is more complicated than the whole of the early universe.[9] He bases this statement on the cosmic background radiation, the afterglow of the Big Bang. When physicists looked for variation in this radiation coming from different directions they found that the temperatures of the early universe, and presumably the density and pressure as well, were uniform to about one part in 100,000. Guth pointed out that, if the tem-perature of the Earth were as consistent as this, a typical weather report would be 'It was scorching today in Dalol, Ethiopia, where the mercury soared to 45.012°F. Meanwhile, at Plateau, Antarctica, there were reports of frigid temperatures as low as 45.001°F.' Weather forecasts would be then very accurate. Why the early universe was not riddled with variations arising from non-linearities is a question with which cosmologists continue to struggle.

This kind of behaviour, in which a system of equations containing no random components exhibits unpredictable results, is *deterministic chaos*. Such a system is called *deterministic* because it is described by equations that determine precisely what will happen following any set of starting conditions. The problem is that the starting conditions cannot be known to sufficient accuracy. As Lorenz realised, exactly where a given storm will pass and when it will drop rain may be extremely sensitive to the details of the winds and temperatures of the atmosphere a few days or even a few hours earlier. The slightest imprecision in the meteorologists' knowledge of these data can quickly render a prediction of the coming weather useless. Because of deterministic chaos it will always be impossible to forecast accurately the weather in a particular area more than a week in advance. The atmospheric linkage between the Gulf Stream and plankton far away across the ocean steps over these kinds of shifting sands.

The term 'chaos' had a much earlier and very different association with the atmosphere. At the time that Joseph Priestley and others were discovering oxygen and other gases, the common term for them was 'airs', since most people regarded them as variants of the common atmosphere. However, some chemists started using the word 'gas', following the earlier physician Jan Baptista Van Helmont. He likened the different vapours he found to unformed matter or pre-matter, and called them 'chaos'. In his heavy Flemish accent this sounded like 'gas', and this is the word that subsequently became accepted.

A large part of the atmospheric problem is turbulence. Any smooth flow in either the atmosphere or the ocean breaks up into whorls and eddies, and energy drains rapidly from the large-scale motions to these smaller structures. The largest of these whorls in the atmosphere are the cyclones and anticyclones with which we are all familiar, but the swirls carry on down to smaller and smaller sizes. This is turbulence. There is a story that the physicist Werner Heisenberg, who won a Nobel Prize for discovering the quantum uncertainty principle, planned to ask God two questions when he died: 'Why relativity?' and 'Why turbulence?' Having worked on both topics, Heisenberg only expected to get an answer to the first question. This story, though probably apocryphal, illustrates how difficult it is to understand the dynamics of fluids.

Unbeknown to Lorenz, the problems of deterministic chaos had already been encountered over 50 years earlier by a French mathematician, Jules Henri Poincaré. Eric Temple Bell, in his *Men of Mathematics*, described Poincaré as 'The Last Universalist'.[10] He may well have been the last man to take virtually all branches of mathematics as his province. It is now generally accepted that it would be impossible for any person starting today to understand comprehensively—much less do creative work of high quality—in the separate fields of mathematics, astronomy and mathematical physics, but this is what Poincaré did. One of Poincaré's achievements was developing the special theory of relativity at the same time as Albert Einstein. Even though he always recognised Poincaré's contribution, Einstein still got almost all the credit for the theory, perhaps because Poincaré was more concerned with the mathematics than the physical meaning of the theory. Poincaré never acknowledged Einstein's work and often referred to the theory as Lorentz's theory of relativity (after the Dutch Nobel Prize winning physicist H.A. Lorentz).

In addition, once his greatness as a mathematician was established, Poincaré began writing books designed to share with non-professionals the meaning and human importance of his subject. These books were both successful and very popular. Reading one of them, *Science and Method*, in my youth, I was particularly impressed by a famous passage in which he described how one of his greatest inspirations came to him. Poincaré described his own experience in developing the theory of a particular kind of mathematical function. He had worked on the problem for 15 days without success. One evening, contrary to his custom, he drank black coffee and then could not sleep: 'ideas swarmed up in clouds; I sensed them clashing, to put it so, a pair would hook together to form a stable combination.' By morning he had the first part of the solution. A day or so later he was boarding a bus that was to take him and some colleagues on a geological field trip. The journey had caused him to forget his mathematical labours but, 'The instant I put my foot on the step the idea came to me, apparently with nothing in my previous thoughts having prepared me for it.' He could not verify the correctness of the solution at the time, because once on the bus he resumed an interrupted conversation, but he still felt complete certainty. On returning to where he was staying, he was able to verify the result. This chain of events illustrates the four stages that people often go through in conceiving an idea: saturation (working on the problem), incubation, illumination, and bringing it to fruition. These four stages in problem solving are now often referred to by psychologists.

Poincaré's investigation of non-linear dynamics arose in part when he was attempting to determine the conditions under which a system of planets or stars will travel in repeating and periodic orbits. This question led him to consider the starting conditions and to discover the unpredictability arising from small errors in these initial conditions. He summarised the difficulty in *Science and Method*:

> A very small cause which escapes our notice determines a considerable effect that we cannot fail to see, and then we say that the effect is due to chance. If we knew exactly the laws of nature and the situation of the universe at any moment, we could predict exactly the situation of that same universe at a succeeding moment. But even if it were the case that the natural laws had no longer any secret for us, we could still know the situation approximately. If that enabled us to predict the succeeding situation with the same approximation, that is all we require, and we should say that the phenomenon had been predicted, that it is governed by the laws. But it is not always so; it may happen that small differences in the initial conditions produce very great ones in the final phenomena. A small error in the former will produce an enormous error in the latter. Prediction becomes impossible.[11]

The problem of chaos arises when the gravitational attractions of three or more bodies are considered. It is now known to be widespread in the astronomical orbits of the solar system and elsewhere. In a recent simulation of the planets orbits over the next 5 billion years, by which time the sun will have turned into a red giant, it was found that in some cases Mercury's orbit was pulled into a collision course with Venus and in one simulation there was the possibility that Earth's gravity could one day tear Mars apart.[12] Poincaré's work was largely ignored in the following decades. One mathematician who was interested was George Birkhoff in the USA, who had briefly taught the young Edward Lorenz at MIT.[7] But chaos is not restricted to physical systems.

The intertwined lives of the snowshoe hares and the Canada lynx have illustrated how non-linear phenomena can happen in biological communities; ecosystems such as that of the plankton contain at least as many interactions. It is therefore not surprising to find chaotic behaviour occurring in complex biological systems. There is, however, a basic difference between biological systems and the physical systems that have been described. Physicists deduce the equations for astronomical or atmospheric motions from the fundamental equations of physics. The laws of motion first written down by Isaac Newton are the foundations of the equations used in meteorology. Biologists, on the other hand, have to construct equations that describe quantitatively the varied ways species behave and interact; they don't have a set of universal biological equations to call upon. The mathematics of how two organisms compete, how a population grows or dies when food is scarce, or how a predator seeks out its prey frequently cannot be worked out from first principles. Consequently, constructing suitable equations often involves making some simplifying assumptions.

It is common to model populations in terms of discrete time intervals. In the simplest case, the relationship between the size of a population in one year and that in the next year is captured by a rule—a function. Following a population through time is then a matter of taking a starting figure and gradually stepping forward in time by applying the same function again and again. Such a model can be used to explore the fluctuations in a population in the absence of the vagaries of the weather. During the 1950s ecologists began using one particular equation, the logistic equation (for example, W.E. Ricker compared its predictions with data from Australian fisheries).

In the logistic equation, the population next year P_{next} is calculated from that this year P by the formula $P_{next} = rP(1 - P)$. It has two components. The first rP represents unrestrained growth rising steadily upward, where r is the natural growth rate of the population, expressed as the factor by which the population increases. Limitations to the population, such as by food supply, predation or habitat, are handled by the second term $1 - P$. This keeps the growth within reasonable bounds since as P rises, $1 - P$ falls. One of the first to explore the implications of the logistic equation was Robert May, then at Princeton University but subsequently at the University of Oxford.[13] May is an Australian who first trained as a physicist before becoming an ecologist. For values of the growth parameter r that are not too large, the model settles to a steady state. Thus, if the parameter is 2.7 the population is 0.629. As the parameter is increased so does the population. But suddenly when the growth rate passes 3 the population refuses to settle down to a single value but oscillates between two values in alternate years. When the growth rate is raised still further the population settles down to four values, each returning every four years. This splitting of the population is called *bifurcation*, and as the growth rate is raised further, more and more bifurcations occur. Suddenly, beyond a certain point, the periodicity gives way to chaos, fluctuations that never settle down at all. May has referred to this behaviour as 'The snake in the mathematical grass'.

The logistic equation is, of course, at best a highly idealised representation of how real populations develop. Even so, when May and others looked at real biological systems they saw features reminiscent of its chaotic behaviour. In epidemiology, for example, it was well known that epidemics of such diseases as measles, polio, and rubella tend to come in cycles that may be regular or irregular. Vaccination of a community is normally carried out following some well-defined strategy, which determines when and to whom the treatment will be administered. Model calculations predict that a programme of vaccination can generate large oscillations in the incidence of the disease, and doctors have seen oscillations like these happening in the course of actual vaccination programmes, such as a campaign to wipe out rubella in the UK. These oscillations can be even more extreme: measles epidemics in New York have been shown to occur in the haphazard fashion of deterministic chaos.

Each species in any ecosystem has growth and population limitation components akin to the logistic equation, even though these are bound up with other parts of the community. It therefore seems that all ecological systems contain the seeds of chaos. The fluctuations of the snowshoe hare and Canadian lynx, as recorded by the trappers of the Hudson's Bay Company, seem to suggest this possibility. While the ups and downs in the numbers of pelts might appear at first sight to be repeating oscillations, closer examination shows them not to be regular either in the times at which the peaks occur or in the maximum sizes reached. If these fluctuations are indeed generated purely by the interactions of the animals, they seem to contain some random element.

This randomness is even more evident in the lives of lemmings, rodents that have periodic population booms and then disperse in all directions, seeking the food and shelter that their natural habitat can no longer provide. When the Bible was translated into Norwegian, mentions of swarms of locusts were accompanied by marginal references to lemmings, for Norwegians knew nothing of locusts but were all too familiar with the episodic explosions in the numbers of these rodents. Across northern Norway in 1970, lemmings were so common that snowploughs were used to clear the squashed animals from the roads. In the 16th century Zeigler of Strasbourg proposed that the sudden explosion in numbers was because the animals fell out of the sky during a storm, before dying when the grass grew in the spring. A more recent myth is that lemmings commit mass suicide when they migrate. In reality lemmings reproduce so quickly that their population fluctuations are chaotic, numbers peaking roughly every four years before they plummet to near extinction. Driven by strong biological urges, they will migrate in large groups when population density becomes too great. Lemmings can swim and may choose to cross a body of water in search of a new habitat. They may also be pushed into the sea as more and more lemmings arrive at the shore, whereupon they swim, sometimes to exhaustion and death.[14]

Chaos therefore can occur in ecosystems. Furthermore, if this is the case, then complex ecosystems should be more chaotic than simple ones, for more complicated systems are likely to include a greater number of non-linear processes. Is it possible that many fluctuations in natural ecosystems are due to deterministic chaos and some of the plankton changes that appeared to be connected to the Gulf Stream could be chaotic? The answer to these questions hinges on how the multiple interconnections in large communities affects their stability.

Whether complex ecosystems are less or more stable than simple ones has long been a consideration of naturalists, and now, as human activities are in danger of reducing the diversity of wildlife, it is becoming an increasingly important question. Ecologists have often suggested that where many kinds of plants and animals live together there will be a better balance than where there are only a few kinds; in other words complexity leads to stability. Paul Colinvaux has illustrated this argument by an extreme example.[15] Suppose there is an island with only two kinds of animals: foxes and the rabbits they eat. If now the rabbits' population was greatly reduced by some accident, the fox numbers would also crash, but before doing so they would be in danger of hunting out the remaining rabbits. If instead the natural accident happened to the foxes, the rabbit numbers might grow explosively until the foxes had bred sufficiently to control them, by which time there might be too many rabbits and the system would take a long time to settle down again. However, if the rabbits lived alongside several different kinds of rodents and the foxes lived in competition with other predators such as weasels and cats, then a drop in the numbers of any one species would not matter so much, some of the others would fill the void.

These kinds of argument suggest that chaotic behaviour might be less likely in a complicated community. But such stability seems to contradict theoretical expectations. When random food webs of different sizes are constructed, Robert May has shown that the stability of the ecosystems decrease rapidly as they become more complex.[16] This implies that real communities are likely to be very vulnerable to chaotic behaviour. To test this, May collaborated with M.P. Hassell and J.H. Lawton in a search for chaotic behaviour in wild insect populations. They found no evidence of it in any field populations they examined.[17] David Tilman, Peter Reich and Johannes Knops at the University of Minnesota have carried out a long-term experimental field test of how diversity affects stability in a grassland ecosystem.[18] During a decade of observations on systems in which the number of perennial prairie species were directly controlled, the abundances of the plant species and their rates of growth varied considerably from year to year because of fluctuations in the climate during the growing season. However, when the data were collated, it was found that greater numbers of plant species led to the plant production being more stable from one year to the next. Over the decade, the stability of the ecosystem with time was significantly greater at higher plant diversity and tended to increase as the different plots matured.

These observations seem to show that chaotic fluctuations may not often occur in biological communities. However, it is still the case that ecological systems often behave unpredictably; the insects and grassland plants just did not fit the mathematical model of deterministic chaos. Ecological food webs certainly have the potential for chaotic responses. When Elisa Beninca and colleagues from the University of Amsterdam cultured a planktonic ecosystem from the Baltic Sea over eight years in a mesocosm that allowed them to keep light and other external factors constant, they found striking fluctuations over several orders of magnitude, which did not repeat but which could be attributed to different species-interactions in the food web.[19] They concluded that predictability was limited to 15–30 days, only slightly longer than the local weather forecast. The planktonic system in the North Sea therefore may well possess the seeds of chaos.

However, the only supporting evidence from ecosystems in the wild for the notion that population fluctuations in nature might be due to deterministic chaos has come from the Hudson Bay pelts and the observations of rodents such as lemmings. A.A. Berryman and J.A. Millstein from Washington State University decided that one way to find out whether the variability in an ecosystem is due to deterministic chaos or to environmental disturbances would be to displace the system from equilibrium and observe its subsequent changes while keeping the environment constant. If the ecosystem were chaotic the displacement would send it off in an unpredictable direction. Unfortunately, natural communities cannot readily be observed in unchanging environments. Berryman and Millstein decided the next best alternative was to analyse the stability of the steady states in mathematical models that have been derived from real ecosystems. When they carried out a survey of modelled systems where this has been done, they found that all exhibited stable steady states.[20] In a later investigation a colleague, Ian Joint, and I examined the steady states of the ecosystem in the upper waters of the Celtic Sea, situated to the south of Ireland. Once again these were stable.[21] Berryman and Millstein produced other evidence that ecosystems do not always behave chaotically. For example, if the unpredictability of ecosystems is due to their deterministic structure rather than weather variations, then we would expect to observe chaos with equal frequency in equable and variable environments. In tropical regions such as the rain forests the climate is less variable and the ecosystems tend to be more complex than in temperate regions, but even so the complex ecosystems are generally more predictable in these tropical regions.

The populations tracking the position of the Gulf Stream must also be non-chaotic. These linkages exist because the communities are driven by the weather patterns, but the relationships would not appear if the ecosystem responded to the weather in an unpredictable manner. How is it possible that intricate food webs are stable when it seems that any disturbance might lead to all kinds of chain reactions among the populations? The answer seems to be that, while in the random food webs the strengths of interactions can assume any values, the rates of processes in real ecosystems seldom, if ever, fall in ranges of values that will generate chaos. This has become known as *May's paradox*. One possible explanation for the paradox is that natural selection has favoured parameter values that minimise the likelihood of extinction. Berryman and Millstein pointed out that the logistic equation predicts that deeper into the chaotic region a population will spend more and more of its time at extremely low densities. But once any population falls close to its threshold for survival, the probability of extinction increases rapidly. In this way, chaotic ecosystems might be weeded out. After all, during the millions of years any species has been in existence it is likely at some time to have experienced one or more decades of continuously unfavourable conditions. The details of how natural selection might have achieved such a weeding out, however, remain unclear and controversial.

One kind of stable community structure was discovered from a long-term study of below-ground food webs in sandy dune soils on a Wadden-sea Island to the north of the Netherlands, and in the central Netherlands. Anje-Margriet Neutal and colleagues from several Dutch institutes found that more biomass at the top of the food chain limited stability, whereas ecosystems with most of the biomass in the levels being consumed were more stable.[22] That is, the most stable food chains had a pyramid shape in which predators and herbivores all had plenty of food available.

The annual progression of the seasons may be another reason that chaotic behaviour is not more prevalent in ecosystems. Every winter, growth rates are reduced to very low values or cease altogether. Furthermore, the means by which a species survives the winter may be very different from what it does in the growing phase; for example, it may lie dormant or it may subsist on a completely different food source. At the end of the winter, populations are often reduced to almost the same abundance each year. Even so, the possibility of chaos is still trapped in ecosystems like a tightly compressed spring and may be released if the conditions become appropriate. The connection between complexity and stability cannot therefore be guaranteed.

This idea of increased complexity generating stability has even been used on a much larger scale, that of an ecosystem encompassing the whole of the Earth. James Lovelock is a scientist who became renowned for pioneering the measurement of trace gases in the atmosphere and who, for many years, has worked as an independent investigator in Devon, UK. Several decades ago, he proposed the Gaia hypothesis. Gaia, named after the ancient Greek goddess of the earth, is used to refer to the system consisting of all living things and all the environments that life affects—soils, atmosphere, oceans, and surface rocks. According to the hypothesis, Gaia is a closely coupled system in which the living things on the Earth actively keep the conditions on the planet favourable for whatever is the current ensemble of organisms.[23] It is a dynamic property of Gaia that the whole of the Earth's ecosystems automatically maintain stability.

To illustrate how this might work, Lovelock considered a world populated by nothing but two species of plant: light- and dark-coloured daisies. This world was supposed to be well-watered, amply supplied with nutrients and having a climate uncomplicated by clouds. When the radiation of the sun shining on this world was relatively weak the dark daisies prospered compared to the light because the dark surface absorbs light better. Absorption of this radiation has the effect of warming up the world. But if the sun's radiation increases, the dark daisies will get too warm for their comfort, and they will do less well than the light daisies which reflect more heat away. Light daisies will then displace the dark plants and this will mean the world absorbs less heat thereby cooling it. The end result is that the temperature of the world is almost unaffected by the strength of the radiation from the sun, at least until one kind of daisy completely covers the world's surface. This 'daisy-world' demonstrates how self-regulation could be a property of a planetary system and result solely from the tight coupling of biological and physical dynamics.

Daisy-world is a very idealised model and so it is a huge step from showing that this kind of stabilisation is possible in principle to demonstrating that the Earth's ecosystems are actually in this kind of balance.

The planet's communities vary widely between seasons: does the Gaia hypothesis apply in winter, summer, or to some kind of annual average. What the idea has done is to reveal the restorative effects, the negative feedbacks, caused by what organisms do to their environment. Whether for their own benefit or not, ecosystems do mould their surroundings, and the Gaia hypothesis has helped expose some of these details.

A good example of such a negative feedback process is the production of dimethylsulphide (DMS) by phytoplankton, especially coccolithophores such as *Emiliania huxleyi*. These algae were described in the last chapter as having extensive blooms that are readily seen from orbit because the white of their chalky shells shows up in the ocean. This gas, DMS, is a breakdown product of the compound that algae use to maintain their osmotic balance with seawater. It is also the gas which gives the smell of seaweed at the seashore. In the sea, algae have to continuously struggle to prevent water seeping from the cells into the more concentrated solution that surrounds them. (In freshwater, the battle is to stop water coming in). This continuous seepage is called osmosis. Some of the DMS that the algae produce leaks out and escapes into the atmosphere, where much of it is converted into sulphate particles which may be a major factor in the formation of clouds over the oceans: in areas far from land these particles provide a large proportion of the nuclei on which water from the atmosphere condenses to form clouds. This process therefore has the potential to contribute to the stability hypothesised by Gaia: increases in solar radiation will raise phytoplankton growth and DMS production, which will in turn lead to more clouds and the blocking out of the sun's radiation. This regulatory chain of processes has been called the CLAW hypothesis after the initials of its proponents.[24]

Some of the CLAW processes have been observed in operation.[25] When Graham Jones and Anne Trevena of Australia's Southern Cross University measured DMS concentrations in corals of the Great Barrier Reef and in its surrounding water, they found that the mucous exuded by the coral contained the highest concentrations of DMS recorded from any organism. A layer rich in DMS formed at the sea surface where it was picked up by the wind. Although globally the emission of DMS from the Great Barrier Reef is not huge, the coral is a concentrated source of DMS which could affect the formation of clouds in the region. Surveys in the 1970s found very high concentrations of aerosol particles in the air above the Reef but, although the coral was thought to be the source, the mechanism by which the Reef might have caused the rise in aerosol particles was not known at the time. Jones and his team, in laboratory experiments, showed that corals produce more DMS when the symbiotic bacteria inside their tissues become stressed by high temperatures or UV radiation. If this DMS seeds clouds, the coral could reduce the water temperature or UV exposure and thus make its environment more comfortable.

However, while these kinds of processes reveal how biological populations influence the earth's environment, they are not sufficient to provide the global stability required by the Gaia theory. Because of the action of winds, the site where the clouds eventually form may be some distance from where the large localised blooms producing DMS occur. Also, in temperate latitudes the phytoplankton grow only in the warm seasons and cannot be effective during winter. Even so, major achievements of the Gaia theory have been highlighting specific processes such as these, and drawing attention to the possibility of stability arising from plants and animals altering their environment.

An unusual example of how living populations change their environment is the way phytoplankton cells adjust their greenness. The green pigment chlorophyll is used by plants to make sugars out of carbon dioxide and water. In the interests of efficiency, the single-celled algae in aquatic systems adjust how much chlorophyll they contain according to environmental conditions, storing more so as to harvest scarce light or making less when nutrients are in short supply. From an analysis of laboratory experiments, my colleague Richard Geider at the UK's University of Essex has shown that these adjustments can be reduced to simple formulae, equations that can then be used to predict how the cells will behave in an actual ocean. Applying these formulae, another co-worker, Nathalie Lefèvre, has shown[26] that the fall in cellular chlorophyll content when nutrients are depleted leads to less light absorption in the mixed layer and cooler sea surface temperatures. At the equator, the temperature reduction is $0.15°C$ throughout the year whereas at $35°N$, a temperature difference of $1°C$ occurs during the summer. This effect is likely to occur over large areas of the ocean and so may be of some global significance.

There are therefore a number of ways in which communities may avoid chaotic behaviour but, even if natural ecosystems do so, they may be driven to chaos by human actions that increase growth rates or induce delays in the regulatory processes. It is always possible to force ecosystems into the chaotic regime by increasing the strength of reinforcing processes (positive feedbacks) within them. Growth rates could be increased through some application of biotechnology, or as a result of stimulating production for economic reasons. We can also create chaotic behaviour by imposing over-compensatory negative feedbacks on the system. An example of this is over-reaction to pest infestations in agricultural systems. Either rapid growth or a sudden crash caused by excessive application of pesticides could generate strong oscillations in a biological community.

Climate change may also shift ecosystems into or out of the chaotic domain. In the case of lemmings, recent warming has led to more stability. Outbreaks of lemming populations in Norway have occurred much less frequently since 1994. Warmer temperatures mean that the snow, in which the animals make their homes, melts and refreezes producing a sheet of ice that they are unable to penetrate either to make nests or to get at the moss on which they feed. There is evidence that the reduction in outbreaks of lemmings has knock-on effects on the wider ecosystem. Scarcity of lemmings means that foxes have been forced to turn their attention to other species such as willow grouse and ptarmigan.[27]

In spite of these man-made possibilities, it seems that in general ecosystems do not gyrate in unpredictable ways caused by their non-linear connections. Instead they tend to track what happens to the weather, thereby showing a more linear response. This gives rise to linkages such as that seen with the distant Gulf Stream. While ecosystems may have the potential for chaotic behaviour without it necessarily being manifested, in the global weather system chaos is always present. In addition, the fluctuations of this system may even mimic those of the logistic equation. When Edward Lorenz explored the dynamics of his simple atmospheric model, and others like it, he found that, just as in the logistic equation, hiding within a particular system could be more than one stable solution. Each such stable solution would represent a different climatic regime. He also discovered yet another kind of behaviour that he called 'almost intransitivity'. Such a system displays one sort of average behaviour for a very long time, fluctuating within certain bounds. Then, for no apparent reason, it shifts into a different sort of behaviour, still fluctuating but producing a different average.

Chaotic fluctuations will always limit day-to-day weather forecasts to no more than about a week ahead. However much meteorological data we assemble, sooner or later some tiny, insignificant movement in some corner of the world's atmosphere will cascade upwards to destroy any prediction. Minute oversights in our observations of the Earth's weather will always prove to be disastrous. But, even though this haphazard behaviour is everywhere, the world's weather still has basic patterns that can be observed: the fluctuations remain tied to a fundamental circulation. Partly, this is due to the long-term memory of the ocean, a memory that arises because the seas can only accumulate and release heat very slowly. Oceans generally have a greater inertia than the atmosphere and this is not just down to heat storage. It has already been shown in Chapter 2 that changes in the Gulf Stream, a major constituent of the North Atlantic climate system, are accumulated from weather events over several years.

Global patterns in the world's weather are the fundamental cause of long-distance linkages such as those seen between the planktonic communities and the Gulf Stream. Therefore, the next chapter discusses the structure of the atmospheric circulation, but it begins with an astronomical event over 200 years ago.

5

EVERY WIND HAS ITS WEATHER

(Francis Bacon, 1561–1626)

The sky has holes for the rain to get in
The holes are small
That's why the rain is thin

(Spike Milligan)

Every now and then the Norwegian fishery turns up a cod with a
deformed skull; the skull has a distinct top or crown giving it the
name 'king cod' or kongetorsk. This rare fish was considered to have the
ability to forecast the weather in days gone by. A woollen thread was
used to hang the fish from the ceiling and its nose would point in different directions depending on the weather to come over the next few days.
In reality, it was not the fish but rather the thread causing the movement.
By absorbing moisture from the air, the twisted thread served as a primitive hygrometer, turning the fish to slightly different positions depending
on the humidity of the air.

One of the first scientists to try to understand weather phenomena with a view to making predictions was Benjamin Franklin. On a
September evening in 1743 he planned to observe an eclipse of the moon,

but storm clouds blew in from the north-east and spoiled his star-gazing. Later he learned that the eclipse was viewed in Boston, hundreds of miles north-east of where he was in Philadelphia, because the storm did not arrive there until four hours after the eclipse. 'This puzzled me', he wrote, 'because the storm began with us so soon as to prevent an observation; and being a north-east storm, I imagined it must have begun rather sooner in places farther to the north-eastward than it did in Philadelphia.' Comparing weather observations that he received in letters from friends in other American colonies and examining newspaper reports from New England to Georgia, Franklin concluded that storms with winds in the north-east started out in the south-west and travelled towards the north-eastern states. In other words, although the winds in a nor'easter blew from the north-east, the storm was actually moving from the south-west at about 160 km per hour. Although today the concept of storms moving from place to place seems obvious, it was not so in Franklin's day. He deduced that the north-east storms arose because of 'some great heat and rarefaction in the air' over the Gulf of Mexico. They were then guided to the north-east by the coast and the Appalachian Mountains.[1]

Franklin even made some of the first-recorded weather forecasts in his Poor Richard's Almanac, a 25-year publication that first appeared in 1732 under the pseudonym of Richard Saunders. Undoubtedly these early observations contributed to his interest in the temperature of the Gulf Stream many years later. In 1763, Franklin took part in discussions about the effect of deforestation on the local climate. Forests were being cleared for farming in the early American colonies and Franklin agreed with other colonial scholars that 'cleared land absorbs more heat and melts snow quicker.' However, he thought that many years of observations would be necessary before any conclusive evidence could be gathered on the effects of deforestation on the local climate.

At the same time that Benjamin Franklin began making the first observations of the Gulf Stream, a young teacher in the English town of Kendal, close to Windermere, was also making meteorological observations. He was John Dalton, but his observations of the atmosphere were ultimately to lead him in a different direction. Dalton was born in 1766 on the edge of the Lake District, near Cockermouth, to a family of poor weavers who were devout Quakers. He was an exceptionally bright student, so much so that at the youthful age of 12 he was put in charge of the local Quaker school. At 15 he took a teaching job in a school at the nearby town of Kendal. In the next century another pupil from this little school, Arthur Eddington, went on to become famous as the astronomer who produced the first observations in support of the general theory of relativity. (It was reported that, when a journalist asked Eddington if it was true that he was one of only three people in the world who understood Einstein's Theory, Eddington considered deeply for a moment before replying, 'I'm trying to think who the third person is.')

Dalton's scientific observations started on 27 March 1787, when he wrote an account of the northern lights, the aurora borealis, in his meteorological journal. He subsequently kept up this journal for 57 years, recording in it a total of 200,000 observations. It is said that his neighbours used to set their clocks by the time he opened his window to read the temperature every morning. In the course of this passion for meteorology, Dalton climbed the majestic 3100-foot Helvellyn 40 times. Part of his interest in meteorology was to understand rainfall and the circulation of water within the atmosphere, perhaps because the Lake District is the rainiest corner of England. In later life, he said that his book of meteorological observations and essays contained the germs of most of the later ideas that he worked on.[2]

Dalton undertook numerous experiments in order to understand how barometric pressure was related to the directions of winds and the heating and cooling of air. One outcome of these experiments was his discovery that, in a mixture of gases, each gas exerts the same pressure as it would if it were on its own; that is the total pressure of a mixture of gases is equal to the sum of their individual pressures. This is Dalton's Law of partial pressures, a law which allows meteorologists to calculate the amount of water vapour in the air and thereby express cloud formation and fog, rain, or snow in mathematical terms. But Dalton's name also appears in a completely different area of science. Once, on his mother's birthday, Dalton bought her some very special stockings. This was meant to be a treat for she always wore homespun stockings, but on seeing them her response was, 'Why did you buy me scarlet stockings?' Dalton had thought they were blue, as did his brother. After his mother confirmed the colour of the stockings with some other women, Dalton realised that he and his brother were both colour-blind and could not distinguish red from blue. In France, the condition of colour-blindness came to be known as Daltonism.

But it was a book called *A New System of Chemical Philosophy*, published in 1808, that really established Dalton's reputation. In it, he argued that all matter is made up of exceedingly tiny, irreducible particles. This idea had been developed by the ancient Greeks but Dalton's contribution was to consider the relative sizes and characters of these atoms and how they fit together. He showed that this could form the basis of chemistry and of much other science. Although this work made Dalton famous, as a Quaker, he tried to avoid all honours. A French chemist, who had travelled to Manchester to visit him, was greatly surprised to find the great man tutoring a young boy. Upon asking if it was Dalton he was talking to, the reply was: 'Yes. Wilt thou sit down whilst I put this lad right about his arithmetic?'

While Dalton, Franklin, and others were making weather observations on land, there were also those who were doing so at sea. When Alister Hardy was making the first deployments of his continuous plankton recorder in the Southern Ocean at the end of 1926, he had a 'little adventure'. At a midnight station, the winch being used suddenly wound some two fathoms of cable instead of six feet in a violent jerk, making the machine leap like a grasshopper before swinging wildly about and crashing into stanchions. An officer and two deck-hands were knocked off their feet, fortunately without injury.[3] It serves a reminder of how easily accidents can happen when operating equipment at sea. Hardy wasn't working on the *Discovery* at this time but on another vessel, the *William Scoresby*. This ship was named after a whaling-captain-turned-scientist whose activities and achievements place him in a class apart from almost all those who have journeyed and researched in the polar regions.

These days it is generally forgotten how large and important the industry of whaling was in the 18th and 19th centuries. Whales were hunted primarily for the oil in their blubber and for the baleen in their mouths. The oil was used to fuel lamps and, in later years, to lubricate machinery. Before the invention of spring steel, elastic and plastic, the flexible baleen was molded and shaped for use in a variety of everyday objects. One notable example was the corsets worn by Victorian ladies. In the mid 1850s, the south-eastern Massachusetts whaling city of New Bedford was one of the wealthiest cities in the USA, even though it had a population of less than 20,000.

In the course of carrying on this most demanding and arduous of all maritime activities with great success, William Scoresby also managed to collect 15 years of data on sea currents and temperatures, ice formation and movement, wind directions and velocities, magnetic variations, marine organisms, the biology of whales, the structure of snow crystals, and much else besides.[4] By lowering a box containing a thermometer into the ocean, he discovered that in the Arctic Ocean the deeper you go the warmer the water tends to get. In his ship's log, Scoresby was one of the first people to use the system of cloud classification recently devised by Luke Howard, essentially the modern description of cloud types.[5] In this, clouds are divided into four basic types: stratus for layered clouds, cumulus for fluffy ones, cirrus for the high feathery clouds associated with colder weather, and nimbus for rain clouds; components can be combined freely to describe every shape of cloud seen, e.g. stratocumulus, cirrostratus, or cumulonimbus. Arctic clouds, Scoresby reported were:

The cirrus, cirro-cumulus, and cirro-stratus, of Howard's nomenclature, are occasionally distinct; the nimbus is partly formed, but never complete; and the grandeur of the cumulus or thunder-cloud is never seen, unless it be on the land...the most common definable cloud seen at sea, is a particular modification, somewhat resembling the cirro-stratus, consisting of large patches of cloud arranged in horizontal strata, and enlightened by the sun on the edge of each stratum.

Scoresby gathered all this original work together in his two-volume classic *Account of the Arctic Regions*, and its publication in 1820 marks the beginning of the scientific study of the poles. In its accounts of whaling adventures, and the dangers and thrills of the chase, the second volume compares favourably with those of the great maritime novelists. When the uncharted coastline of east Greenland was clear of ice around 1820, Scoresby, in the midst of an arduous whaling voyage, sailed along some 400 miles of this inhospitable landscape, charting it and naming points as he went in honour of scientific and other friends, chief of which was Scoresby Sound, named for his father, who had also been a whaler.

A favourite trick of Scoresby Jr was to use the warmth of his hands to mould ice into a lens and then use it to light his sailors' pipes by means of the sun. William Scoresby left active sea life in his thirties to enter the Church in his birthplace of Whitby. Despite a busy life, he still continued to work for science with his pen, sending many papers to the Royal Society. He visited America and Canada twice in the 1840s, lecturing to support his many social endeavours in the industrial parish where he pioneered five schools for the illiterate mill-working children of Bradford. In 1826 he preached at a service held for the loss in storms of the *Lively* and the *Esk*, the last two whaling ships to sail from Whitby. William Scoresby's lasting legacy is some of the earliest meteorological and oceanographic observations of the North Atlantic and the Arctic.

Franklin, Dalton, Howard, and Scoresby were just the vanguard of a network of weather observers, both on land but also from ships. Widespread measurements formed the foundations on which the science of meteorology would be built, but progress with understanding how the world's weather system works could not be made until all these disparate observations were collected together and synthesised. Perhaps the most significant person in this process was Matthew Fontaine Maury of the US Navy, who systematically compiled a definitive set of wind and current charts in the middle of the 19th century.[2,6] Maury would never have taken on this task but for a piece of tragic misfortune.

Matthew Maury was a lieutenant in the US Navy until an accident ashore crippled him and brought his seagoing career to an end: On 17 October, 1839 he was on a coach travelling to report to his ship, the *Consort*. When at the staging post in Lancaster, Ohio there were three more passengers waiting to board the already overcrowded vehicle, he gave up his inside seat to squeeze on top. The road was stony and winding and, top heavy with extra passengers and excess baggage, the coach toppled over. Thrown from his perch, Maury landed on his bent leg, fracturing his thigh and dislocating his knee. The *Consort*, sailed without him while he spent several weeks lying in an inn unable to move.

Maury was too crippled to go to sea again and so the Navy put him in charge of the Depot of Charts and Instruments, which contained the charts and logs of every voyage undertaken by a naval vessel. Maury realised that there was a wealth of information to be discovered about ocean currents and weather by analysing the data in the logs, and so he began plotting the records on blank charts. It soon became obvious that the weather systems around the world were connected, and Maury proposed some of the first speculative theories of how the oceans and atmosphere circulated. He quickly realised that there was a need to incorporate observations from outside the USA, but this raised a problem: as there were no agreed standards for either the instruments or the methods of observation, direct comparisons could not be made between reports from American and foreign ships. To remedy this, in 1853, Maury called the first international meeting of meteorologists. He managed to get ten nations to take part. One of these was Britain whose delegation was sent with (perhaps typically) strict instructions not to involve the government in the expenditure of any money. Maury's meeting determined a fixed standard for the making and recording of weather observations from ships at sea, and set up a system for the friendly exchange of such information between the countries.

While the British government wasn't prepared to donate money to an international project, following the meeting, they did decide to set up their own organisation, the Meteorological Department of the Board of Trade. Admiral Robert FitzRoy was appointed to be its director. FitzRoy is best known today as the captain of the naval ship HMS *Beagle* during the voyage that took Charles Darwin round the world and led him to propose the theory of evolution by natural selection. Darwin had been invited on the voyage primarily as a dinner companion for the captain, whose rank precluded his socialising with anyone other than a gentleman. FitzRoy supposedly chose Darwin partly because he believed the shape of Darwin's nose betokened depth of character. Robert FitzRoy was a deeply religious man and, when the theory of evolution was published 25 years later, bitterly regretted taking Darwin on the voyage. When the theory was debated at a famous meeting of the British Association for the Advancement of Science in Oxford, FitzRoy walked through the hall holding a Bible aloft and shouting, 'The Book, the Book!' FitzRoy suffered from mental problems which ultimately led to his committing suicide using a razor.

Robert FitzRoy was at the conference to present a paper on storms, in his capacity as head of the Meteorological Department,[7] a paper that was totally overshadowed by the discussion of Darwin's work. FitzRoy's passionate concern was for the safety of ships at sea. Hundreds of lives were lost every year in storms, and he wanted to do something about it. He realised that by using the telegraph, which had been invented in 1838, he could discover very quickly what the weather was doing in many different places at the same time, so he set up weather stations around the coast and plotted the data onto charts, which he said would provide 'a synoptic view of the weather'. 'Synoptic' refers to observations all taken at the same time. From his charts, FitzRoy could see weather systems forming and moving and so was able to issue storm and gale warnings for ports and harbours around the country. A system of cones suspended from flagpoles was devised to warn ships of impending bad weather. A solid black cone pointing downwards warned that a southerly gale was imminent, while a cone pointing up warned of a northerly gale. These cones were used in ports around the British Isles for more than 100 years.

FitzRoy also started publishing forecasts in newspapers such as *The Times*, and Queen Victoria was known to have consulted him occasionally about the weather. However, these experimental forecasts were often wrong and were criticised, with the result that *The Times* ceased publishing them shortly before he died. After his untimely death in 1866, the Meteorological Department was closed, but the uproar from the navy and merchant marine at the loss of their storm warnings saw it opened again, and FitzRoy's work was continued. FitzRoy and his half-brother Charles, who was Governor of New South Wales for a time, are remembered in Australia by the rivers named after them. The Fitzroy River in Queensland is named after Charles, while Robert has the beautiful Fitzroy River in Kimberley, Western Australia.[8]

In making his forecasts, FitzRoy, like Franklin before him, was tracking weather systems. Undoubtedly, the most familiar of these, now seen daily on TV and in the newspapers, are cyclones and anticyclones, usually expressing 'bad' weather (cyclones) or 'good' weather (anticyclones). Cyclones are areas of low pressure around which the winds blow anticlockwise in the northern hemisphere and clockwise in the southern hemisphere. Anticyclones are areas of high pressure around which the winds blow clockwise in the northern hemisphere and anticlockwise in the southern hemisphere. The direction of winds in both cyclones and anticyclones is another manifestation of the rotation of the Earth. In the tropics, cyclones originate within a few degrees of the equator in the eastern half of the Atlantic and Pacific Ocean before moving westwards as they gather in strength. Further north, these highs and lows account for the variable day-to-day weather in the northern temperate zone as they move around the globe from west to east.

Out of these and other initial observational threads was gradually woven the pattern of global atmospheric circulation which is summa-

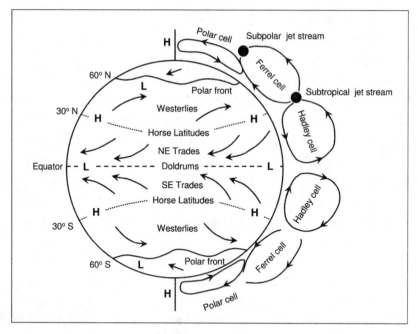

FIG 5.1 A summary of the world's weather patterns.

rised in Fig. 5.1. The first attempt to make sense of the world's weather patterns was made in 1735 by George Hadley, a lawyer with a keen interest in the weather[9] (George's brother John was the inventor of the octant, a precursor to the modern sextant). The world's weather exhibits certain gross patterns: hot and wet in equatorial areas (up to 10° latitude either side of the equator), hot and dry between 10° and 35° latitude, cool and wet between 35° and 75° latitude, and predominantly cold and dry poleward of 75°. Within the tropics, the trade winds blow westwards between the equator and a latitude of about 30°. The trade winds are regular in both strength and direction and, as such, are very different from the winds farther away from the equator. Both calms and gales are rare in the trade-wind belts. Instead, light to moderate breeze, with clear sunny weather continue day after day. As they were vital for European sailing vessels travelling to the shores of North America, understanding the trade winds was important in Hadley's day.

Between 30° and about 70° in each hemisphere (a band that contains the British Isles and the North Sea) the winds are mainly westerly (that is, they blow from the west). As described in Chapter 2, this wind gyre of easterly wind at low latitudes and westerly winds further north drives a similar gyre in currents of the North Atlantic Ocean, part of which is the Gulf Stream. This gyre has been shown in Figs 2.2 and 2.3, and schematically in Fig. 2.4. Similar gyres occur in other oceans both north and south of the equator. Superimposed on this simple pattern of easterly and westerly winds are some variations caused by differences in the distribution of land and sea, and the progression of the seasons, but Hadley sought to isolate a basic underlying pattern.

Building on an earlier theory by Edmund Halley (after whom the comet is named), Hadley realised that the heat from the sun is unevenly distributed, being greater at the equator than the poles. He argued that this unevenness causes differences in air pressure and then air flows from areas of high pressure to areas of low pressure to restore equality. These movements of air are the winds. High temperatures at the equator cause the air at the surface to expand, become lighter and rise upward. It is replaced by air flowing in from north and south, which is cooler and therefore heavier. The trade winds are simply this heavier air moving over the surface. It might be expected that they would be north or south winds. However, the Earth is spinning from west to east and so, to anyone on the planet's surface, the trade winds appear to come from the north-east in the northern hemisphere and from the south-east in the southern hemisphere (this is the Coriolis effect described previously for currents in the ocean). Hadley proposed that a 'cell' existed between 0° and 90° in each hemisphere with warm air rising over the equator, moving poleward at high levels, sinking at the poles and returning equatorward at the surface. Later observations showed that this does not happen, but a cell operating in the sense that Hadley described does exist between 0° and 30° in each hemisphere. The equatorial low pressure region between these two cells, an area of light and variable winds, is called 'the doldrums'.

The Hadley cell is a vertical overturning system: warm air rises, cold air sinks and two horizontal flows complete the circulation. Near the equator, warm, moist, unstable air, largely over the ocean, rises in a multitude of cumulus towers. When condensation of the water vapour occurs at high altitude the heat released warms the upper tropical atmosphere. At heights of 12–15 km the air is cooled by radiating heat, a process that continues as the pressure gradient drives the air poleward. At latitudes of about 30° north or south the air sinks in the subtropical high pressure areas, warming by compression (rather like a bicycle pump warms when used vigorously) and inhibiting cloud formation. Hence skies are frequently clear, and the incoming solar radiation and surface temperatures are both high. Here lie the hot, dry deserts of the world such as the Sahara and the Australian Desert. In the ocean, these regions are sometimes called 'the horse latitudes'. The term supposedly originated because Spanish ships transporting horses to the West Indies sometimes ran out of water and their crews had to throw horses overboard when they became becalmed in mid ocean. This story may be apocryphal, however.

While a Hadley cell does correspond approximately to part of the atmospheric circulation, it is still a simplification of a more complicated system which differs over land and sea. Over the oceans, two particular areas of semi-permanent high pressure exist in the subtropics: a high-pressure centre over the Pacific westward of California and another over the Atlantic, near the Azores and off the coast of Africa. Pressure is also high, but less persistently so, west of the Azores to the vicinity of Bermuda (Fig. 5.2). These subtropical highs are more intense, and cover a greater area and extend farther northward in summer than in winter. In the southern hemisphere there is also a belt of high pressure roughly parallel to 30°S, which has three centres of maximum pressure over the oceans. One is in the eastern Pacific, the second in the eastern Atlantic, and the third is in the Indian Ocean.

Hadley came to realise that there was more than one circulation loop at work. Apart from the tropical Hadley cell there was a polar Hadley cell in which cold, heavy air from the north flows southwards to warmer areas, gets heated up by the ocean surface and then rises and returns northward. While Halley's earlier theory was widely known, Hadley's remained unknown for about a century, although it was rediscovered several times. One of these later discoverers was John Dalton, who subsequently acknowledged Hadley's priority.

In 1856, the American meteorologist William Ferrel proposed that between tropical and polar cells in each hemisphere there is a third cell. Because of the Coriolis effect, the cold, dense air returning to lower latitudes in the polar Hadley cell does so as easterly winds. At about 50–60°N and 50–60°S of the equator, the surface winds meet and the air rises. At height, the air divides, with some air flowing to the poles and some towards the equator forming the Ferrel cell. The Ferrel cell produces westerly winds between the tropical and polar Hadley cells. Ferrel's model indicates where rising air will lead to significant precipitation, and where sinking air will create high-pressure zones with arid conditions, the hot deserts. The model encapsulates the law proposed by Hendryk Buys-Ballot in 1897: if in the northern hemisphere you turn your back on the wind then low pressure is on your left and high pressure is on your right. In the southern hemisphere it's the opposite way round.[1]

Another weather pattern driven by upward and downward movements of air is the Asian monsoon. Its name comes from the Arabic *mausin*, meaning a season, which is apt as monsoons are circulations that reverse their direction every six months. The monsoon is somewhat like a gigantic sea breeze. In summer, the Asian continent heats up more than the nearby ocean which leads to high pressures at high levels over land as the warm air rises from the land. Air pressure at this height over the sea is lower and so air moves from land to sea, creating an opposite pressure gradient at the surface. This drives air from sea to land at the surface.

Although widespread across Asia, the monsoonal circulation is especially well developed over the Indian subcontinent, which juts out into the Indian Ocean. The country's rice crop depends on the moisture it brings. Monsoons have always played a vital role in the economy of the

Middle and Far East. Near-surface flow sweeps over southern Asia from the southern hemisphere, and close to India these winds bring heavy rains and then swing around to become south-easterly up the Ganges valley. The heaviest rainfall is over the Assam hills, as the winds approach the Himalayas. Above these low-level winds, tropical easterlies flow from the western Pacific to Africa over a band of about 15° across the equator.

In winter, the circulation is reversed because the land cools down much more than the sea in this season. The low-level winds are north-easterly over India and north-westerly over Chinese Asia. Cold, dry air flows south-westward from the Asian interior giving relatively dry conditions. In the upper atmosphere, westerlies now overlie all but the extreme southern part of the monsoon regime.

Outside the tropics, steady flows of air, such as are experienced in the trade winds and the regular swaying back and forth of the monsoons, are absent. Here, the vertical overturning becomes unstable and collapses into a system of waves, called *Rossby waves*, which progress in the upper atmosphere around the globe, connecting up weather patterns over great distances. These motions of the upper atmosphere are named after Carl-Gustaf Rossby, the eminent Swedish-American meteorologist who provided the first theory of them. The waves are north–south oscillations stretching around the pole and at their shortest they are several thousand kilometres long. They extend a similar distance in the north–south direction, with crest-to-trough distances of over tens of degrees of latitude.

These waves in the upper atmosphere are linked to the movements of weather systems beneath them, steering cyclones on their journeys across the Earth's surface. Their progression around the globe is caused by the planet's rotation. Just as happens in the ocean, the spin associated with any motion of the air when added to its rotation about the Earth's axis (i.e. its total vorticity) must remain constant. A stream of air moving towards polar regions will tend to gain vorticity as the distance from the Earth's axis of rotation reduces. This anticlockwise vorticity has to be compensated by the flow adopting an anticyclonic curvature (clockwise in the northern hemisphere) relative to the surface. Eventually, the curved path takes the airstream back towards the equator. When the stream of air reaches lower latitudes, the reverse effect will take place, as the distance from the axis of rotation increases, and the airflow will adopt a cyclonic curvature. So the airstream tends to swing back and forth as it circulates the Earth.

The waves tend to originate where there are massive mountain ranges such as the Tibetan plateau and the Rocky Mountains, but the distribution of land and sea may also reinforce the pattern. In the northern hemisphere they frequently have major troughs at around 70°W and 180°W, whose positions are linked to the major mountain ranges, together with effects of heat coming from the ocean currents in winter and from the land masses in summer. The southern hemisphere is largely covered by water and so the pattern is much more symmetrical and shows less variation from winter to summer.

Rossby waves take on a myriad of shapes but two extremes are particularly important. When the poleward temperature gradient is strong, the waves tend to have small amplitudes, long wavelengths and to contain relatively high wind speeds. North–south temperature and pressure gradients across the waves are strong, which encourages the rapid easterly movement of depressions with effects on regions in their path. When the latitudinal gradients are weak, there is an increase in both the number of swayings back and forth and their amplitude. In turn, these lead to 'blocking', where high-pressure systems act rather like boulders in a stream. This causes extremes of weather, particularly in north-west Europe. In summer, high temperatures can occur for several weeks, such as happened in the dry summer of 1976, while, in winter, very low temperatures, ice and snow may last for months as in the harsh winter of 1962–63. The atmosphere over the North Atlantic Ocean has a recurring tendency to fluctuate between these two extremes.

Beyond the belt of high pressure, in each hemisphere, the surface atmospheric pressure diminishes toward the poles, a fall that reflects the passage of low-pressure cyclones. Although far less savage than those in the tropics, extratropical cyclones are larger, last longer, occur more often and affect the weather over much larger areas. In the southern hemisphere they occur during all seasons of the year over a continuous belt around the globe between 50 and 60°S, while the corresponding band in the northern hemisphere is centred at 50°N in winter and 60°N in summer. Because there are more land-masses in the north, the belt of low pressure is much less uniform. In the vicinity of Iceland, atmospheric pressure is low most of the time (Fig. 5.2). Possibly this can be attributed to the cold land of the island being surrounded by warmer surface water coming from warm ocean currents which separate it from the icecaps of Greenland. The so-called Icelandic low is most intense in winter, when the greatest temperature contrast occurs, but it persists with less intensity through the summer. Near Alaska, a similar situation exists in the Pacific Ocean with what is called the Aleutian low. This region of low pressure, situated over the Aleutian Islands is most pronounced when the neighbouring areas of Alaska and Siberia are snow covered and colder than the adjacent ocean. It is important to realise that the Iceland and Aleutian lows are statistical rather than real. They are regions where cyclones frequently arrive from elsewhere, not areas where one cyclone persists as a permanent feature. Here the cyclones remain stationary or move sluggishly for a time, before they move on or die out and are replaced by others. Occasionally these regions of low pressure may even be invaded by travelling high-pressure systems.

Superimposed on these broad patterns is a more detailed structure due to the distribution of land and sea, the annual cycle of the seasons, and individual weather events. All these causes are encapsulated in Francis Bacon's quotation at the head of this chapter, summarising the common observation that sunshine, warmth, cold, rain, and other elements are related to the winds and where they come from. Meteorologists find it useful to analyse these sources of the weather in terms of air masses: huge volumes of the atmosphere in which horizontal gradients of temperature and humidity are comparatively small. The existence of air masses was first proposed by the Norwegian meteorologist Vilhelm Bjerknes, one of three generations working in oceanic and atmospheric sciences. His father Carl Anton Bjerknes, professor of mathematics at

Christiania (now Oslo) University, provided data from his laboratory experiments in support of Ekman's theoretical analysis of the effects of the Earth's rotation on ocean currents. Vilhelm's son, Jacob, will appear later as an investigator of the El Niño phenomenon. Carl-Gustav Rossby also began his career working with Vilhelm Bjerknes.

In 1917, Vilhelm Bjerknes set up a series of weather stations throughout Norway to produce general pictures of weather conditions over a wide area. Studying these pictures led him and his colleagues to conclude that there exist various distinct masses of air, and that these masses are separated by definite boundaries. Likening the masses to opposing armies (this was during the First World War), they called the boundaries between them *fronts*. The group went on to develop the theory of the formation and dissolution of these features. Air masses form over large homogeneous surfaces such as ice (in Greenland or Antarctica), tundra and forest (in Canada), or the oceans. As a consequence, they become relatively warm, cold, dry, or humid as the case may be. Tropical air masses originate in low latitudes and polar air masses in high latitudes. Maritime and continental air masses are associated with sea and land, respectively. At any time, it is possible to divide the atmosphere into areas dominated by particular air masses. These give rise to different combinations of weather, such as biting cold, dry air over a continent in winter, or warm muggy air over coastal areas in summer (such as at the time of writing this). Once weather of a certain character has formed in an air mass, it tends to be carried by the wind and exported to other areas. Herein lies the truth of the saying 'Every wind has its weather'.

Fronts, the familiar lines of cloud and rain seen on weather maps, form at the meeting of different air masses. One particular front, where cold polar air meets warm tropical air in temperate latitudes, produces especially dramatic effects. During the Second World War, on 1 November 1944, Captain Ralph D. Steakley flew one of the first B29 US high-altitude reconnaissance missions over Tokyo. He and his crew were suddenly subjected to very heavy winds putting their lives in jeopardy. 'I found myself over Tokyo with a ground speed of about 70 mph. This was quite a shock, particularly since we were under attack from anti-aircraft guns and were a sitting duck for them. Obviously the head wind was about 175 mph.' He did not know it but this was one of the first observations of a *jet stream*. These fast ribbons of wind occur at 10–12 kilometres up in the atmosphere, are 80–600 kilometres wide, and have speeds of up to 400 kilometres per hour in the centre. They blow from west to east, following the polar front.

Ironically, the Japanese already new about the jet stream from research on balloons carried out before the war. At about the same time as Steakley's mission, they began using the jet stream to carry high altitude balloons called *fugos* with the intention of attacking the North American mainland and using incendiary bombs to start fires in the forests of the Pacific north-west. Nine thousand fugos were launched, out of which about a thousand reached America killing six people. As it turned out, the forests were too wet from rain and snow to be set alight. By chance, the most success-ful fugo attack brought down the power line to Hanford—the plutonium plant that was producing the material from which the atomic bombs were being made. Flights across the North Atlantic and Pacific Oceans now routinely take account of the position of the jet streams.

Rossby waves, air masses, fronts and jet streams all conspire to ensure that weather fluctuations are interconnected over geographical distances. Together they steer cyclones in their trajectories across the oceans. Con-sequently, there is ample scope for the winds driving the Gulf Stream to be linked to the weather experienced by distant plankton. However, the observed coupling spanning the ocean needs more than an occasional connection caused by the passage of a particular cyclone; patterns of weather over different years must be connected. There are characteristic ways in which the climate over the whole North Atlantic Ocean tends to fluctuate. Hans Egede Saabye was a Danish missionary who wrote a history of the climate, flora, and fauna of Greenland in the 18th centu-ry.[10] In it, he wrote: 'In Greenland all winters are severe, yet they are not all alike. The Danes have noticed that when the winter in Denmark was severe, as we perceive it, the winter in Greenland in its manner was mild,

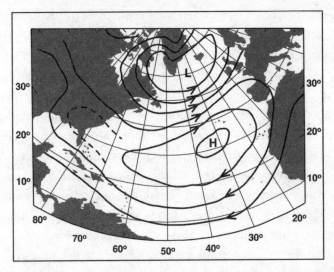

FIG 5.2 The distribution of average atmospheric pressure at the surface of the North Atlantic showing the high pressure over the Azores and the low pressure over Iceland. The arrows show the direction of the winds, which generally follow the isobars.

and conversely.' We now know that he was describing a recurring pattern. Coupled fluctuations in temperatures, rainfall, and sea level pressure have subsequently been documented, reaching eastwards to central Europe, southwards to subtropical West Africa and westward to North America. This pattern is typified by the change from the winter of 1994–5 to that of 1995–6. The latter winter was colder than normal by 2–4°C all the way from Scandinavia to the Black Sea. Rainfall in much of the area was 50% below normal, whereas it was much higher than usual over the western Mediterranean and North Africa. By contrast, the earlier winter was typical of the conditions that had been in place, more or less, since the 1970s with an opposite distribution of weather.

It is now known that what had happened was a swing from a phase with an intense low over Iceland and a strong high over the Azores to a phase in which each of these features was much weaker. This swing was caused by a shift in the number and paths of storm systems crossing the ocean, and it is these changes that were responsible for the reduction in rainfall and drop in temperature. The seesaw behaviour between these two climatic states has become known as the *North Atlantic Oscillation,* or the NAO. It is the dominant pattern seen when atmospheric changes across the North Atlantic are examined over periods of years or decades. The NAO shifts between a deep depression near Iceland and intense high pressure around the Azores, which is accompanied by strong westerly winds, and the reverse pattern with much weaker westerly winds.[11] These two opposite phases of the NAO are also associated with the two extreme Rossby waves described earlier. These shifts are particularly manifest in winter. The strong westerly pattern pushes mild air across Europe and into Russia, while pulling cold air southwards over western Greenland. The strong westerly flow tends to bring mild winters to much of North America, together with a reduction of snow cover, not only during the winter, but well into the spring. The reverse meandering pattern often features a blocking anticyclone over Iceland or Scandinavia which pulls Arctic air down into Europe with mild air being funnelled up towards Greenland. This produces much more extensive continental snow cover reinforcing the cold weather in Scandinavia and eastern Europe, which often extends well into the spring.[12]

This phenomenon is to some extent a coupling between the Atlantic Ocean and the atmosphere. During winters in which there is a steady cooling of sea surface temperature in the band of latitudes from 40 to 50°N, the north–south pressure gradient is found to be strong. Higher than normal winds transfer heat from the ocean to the atmosphere, leading to oceanic cooling. While this cooling is taking place the climate of western Europe is warm and wet. At the other extreme, steady warming of the sea surface over winter, relative to what is usual, is associated with reduced heat transfer from the ocean to the atmosphere.

Fluctuations in the strength of the NAO are connected with many other widely separated physical and biological changes in the region. Since the time of Sir Gilbert Walker, the strength of the NAO has been measured by the difference between the winter atmospheric pressure in the vicinity of the Azores and that in the vicinity of Iceland, a number that has been called the *NAO index*. A high value of the index then corresponds to an intense Azores high and Iceland low and strong westerly winds, while with low values of the index these are all weak, and blocking is common.

The oscillations in winter between high values and low values of the NAO index account for about half the variability in monthly temperatures in central England. But it's not just wind and warmth. Mild winters with a strong NAO dry out the south of Europe, leading to everything from poor olive harvests to poor skiing. Atmospheric circulation patterns force variations in the extent of sea ice and, as a result, the changes in sea-ice cover in both the Labrador and East Greenland Seas as well as over the Arctic are strongly affected by the NAO. A strong NAO brings down more sea ice in the Labrador Sea while compressing that in the Greenland Sea. A weak NAO has the reverse effect. Because of its strong association with temperature, precipitation, and surface winds, the NAO has a strong influence on the biosphere, and so its fluctuations have been accompanied by many biological changes in this sector of the northern hemisphere.

Indeed, some of the observed trends that appear to reflect global warming may instead be due to recent changes in the NAO. A striking feature of the 1990s was the almost unprecedented run of years in which the NAO has been strong, and Jim Hurrell of the National Center for Atmospheric Research in Boulder, Colorado, who is one of the leading investigators of the phenomenon, has shown how these recent trends in the NAO can account for some of the trends in the climate over Europe.[11] The NAO is a globally significant phenomenon and has a powerful influence on northern temperatures. As a consequence, its status has a noticeable effect on the average temperature of the globe.

Another aspect of the North Atlantic that is subject to its whims is the path of the Gulf Stream.[13] In Chapter 2 we saw how the way that the Gulf Stream system integrates the atmospheric forcing over more than a year

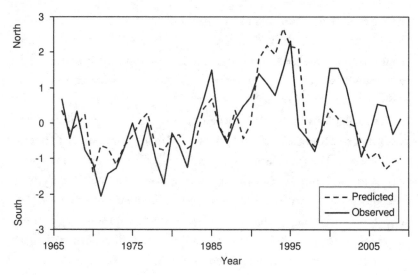

FIG 5.3 Latitude at which the Gulf Stream leaves the US coast as predicted by the Behringer, Regier, and Stommel model compared with observations over 40 years (the graphs are in standardised units).

can be replicated in a simple model in which the ocean is represented by a single line of boxes running from north to south (Fig. 5.3). The wind data used in the model were estimated from the NAO index. Variations in the winds from year to year other than those associated with the historical fluctuations in the NAO therefore were excluded. The model utilises the observed tendency for the wind system to be displaced north and south as the NAO rises or falls in intensity. Movement of water and redistribution of heat within the model then have the effect of integrating the wind-forcing and introducing a delay. Figure 5.3 serves to illustrate how far the ocean circulation in the North Atlantic is at the beck and call of the NAO'.[14,15]

However, the detailed mechanism may be more subtle. Sultan Hameed and Sergey Piontkovski at Stoney Brook University, New York have found that the position of the north wall of the Gulf Stream is much more dependent on the strength of the Iceland low than on that of the Azores high. They suggest that the major oceanic control of the northwall may be southward flow of the cold Labrador current along the northern coast of the USA.[16]

The ups and downs of the NAO must therefore be at the root of the connection between biological communities around the eastern Atlantic and the distant Gulf Stream. But any regional climatic process does not occur in geographic isolation, and so the timing and strength of such cycles almost certainly have their origins in the global climate system. Therefore, it is likely that the NAO is driven by events elsewhere across the planet, but even so it shows trends and decadal variations that indicate that the large heat capacity of the oceans is hindering its variability. One reason for thinking that the NAO is driven to some extent by the exchanges between the ocean and the atmosphere is that this is now known to be happening in other parts of the world. The NAO is not the only large-scale seesaw in the planet's climate, and neither is it the most dramatic. It is but one of several such oscillations going on in different locations. They were discovered in the course of an investigation into a tragedy, a serious problem that turned out to be global. But the man who came across all these climatic seesaws, a bureaucrat in a colony of the British Empire, had little idea how they operated.

6

THE TANGLED SKEINS OF THE WORLD'S WEATHER

'I cannot help believing that we shall gradually find out the physical mechanism by which these (relationships) are maintained...'

(Gilbert T. Walker, 1918)

In 1897 Mahatma Gandhi, who had been born in India, was living and working in South Africa. Yet, he was still concerned with the plight of his homeland:

LETTER TO 'THE NATAL MERCURY'

DURBAN,
February 2, 1897

THE EDITOR, *The Natal Mercury*
SIR,

I venture to offer a few remarks on the Indian famine, regarding which appeal for funds has been made to the British Colonies. It is not perhaps generally known that India is the poorest country in the world, in spite of the fabulous accounts of the riches of her Rajas and Maharajas. The highest Indian authorities state that 'the remaining fifth (i.e., of the population of British India), or 40,000,000, go through life on insufficient food'. This is the normal condition of British India. Famines, as a rule, recur in India every four years. It must not be difficult to imagine what the condition of the people would be at such a time in that poverty-stricken country. Children are snatched from their mothers, wives from their husbands. Whole tracts are devastated, and this in spite of the precautions taken by a most benevolent Government. Of the famines of recent times, that of 1877–78 was the most severe. The famine commissioners thus report as to the death-rate:

It has been estimated, and, in our opinion, on substantial grounds, that the mortality which occurred in the Provinces under British administration, during the period of famine and drought extending over the years 1877 and 1878, amounted, in a population of 197,000,000 to 52,50,000 [sic.] in excess of the deaths that would have occurred had the seasons been ordinarily healthy.

The total expenditure during the crisis was over £11,000,000. The present famine bids fair to beat the record in point of severity. The distress has already become acute. The worst time has yet to come, when summer sets in. This is the first time, I believe, that the British Colonies have been appealed to from India…

The Central Famine Committee at Calcutta must have exhausted all the resources before deciding to appeal to the Colonies. And it will be a great pity if the response is not adequate to the urgency of the appeal. It is true that the outlook is not particularly cheerful even in South Africa, but it will be admitted that there can be no comparison between the distress in India and that in South Africa. And even if there should be a call on the purse of the Natal magnates on behalf of the South African poor, I venture to trust that that would not deter them from dipping their hands deep into their purses on behalf of millions of their fellow-subjects in India, who are on the verge of starvation. Whether it be in the United Kingdom or in the Colonies, I am sure British philanthropy will assert itself, as it has on previous occasions, on behalf of suffering humanity, no matter where and how often.

I am, etc.,

M. K. GANDHI
The Natal Mercury, 4–2–1897

Gandhi was trying to draw South Africa's attention to the currently growing drought in India by warning that there was a danger of it developing into the severity of the situation 20 years earlier. Although it did get worse, it still never became quite as terrible and tragic as that in the years 1877–8. The drought in these years was all over Asia. In China it was known as the Great Famine. For two years, while the south of China suffered a series of devastating floods, hardly a drop of rain fell in northern and central regions. The Shanxi Governor, Tseng Kuochuan, described how: 'The sun glitters red in mid-heaven and its scorching terrors blaze abroad the earth. The seed-time having gone by, and the rain having failed, it is useless to think of planting.' By the time the rains of the 1879 summer monsoon arrived, between 9 and 13 million people had died, with another 70 million severely affected. In some counties of the north-east, up to 90% of the population disappeared. At the time, it was the world's worst recorded natural disaster.[1]

Gandhi's letter describes how disastrous the Great Famine was in India. The drought began in 1876 with a partial failure of the rains in the south of the country, becoming a nationwide calamity in 1877 when the annual summer monsoon failed in large parts of the north, south, and north-western provinces. The winter that followed was extremely severe, killing many from exposure and bringing destructive hail storms in February. Although parts of southern India received rains in 1878, most of the country remained parched and the resulting famine was one of the worst in India's history: by 1879 over five and a half million people were dead. But the Chinese and Indian famines were not isolated events, for weather disasters were occurring around the globe. North-east Brazil suffered a drought unsurpassed in its severity, while much of the USA's Midwest and Missouri Valley basked in unusually high temperatures, forcing many farmers to harvest by moonlight. Even so, in human terms these weather events were most tragic in India and China.

India was part of the British Empire at this time and the British Government's response to the disaster was to establish an observatory with the aim of predicting fluctuations in the Asian monsoons, and thereby preventing future famines. Seeking possible causes of the phenomenon, the first director general of this Indian Meteorological Department, Henry Blanford, turned to a study of conditions beyond India's shores, drawing on information from Britain's other colonies. In the course of these enquiries, he obtained what later turned out to be the crucial reply from the South Australian government astronomer, Charles Todd. This letter contains the first definitive recognition of an international climatic connection:

'Comparing our records with those of India, I find a close correspondence or similarity of seasons with regard to the prevalence of drought, and there can be little or no doubt that severe droughts occur as a rule simultaneously over the two countries.'

But Blanford missed this point, instead basing his attempts at forecasting monsoons on the hypothesis that 'varying extent and thickness of the Himalayan snows exercise a great and prolonged influence on the climate conditions and weather of the plains of north-west India'. The significance of Todd's observation was not fully appreciated until Sir Gilbert Walker, a British mathematician, entered British Colonial Service as director general of the observatory in 1904.

Walker initiated extensive statistical studies of how weather elements such as pressure, temperature, rainfall, etc. varied across the world, with the aim of developing a systematic method for forecasting monsoon rainfall over India. In this, he was expanding on earlier work by the Swedish meteorologist Hugo Hildebrand Hildebrandsson, who had drawn attention to the global scale of oscillations in surface atmospheric pressure. Walker's approach was to use correlation coefficients to measure how similar were the changes occurring in two locations. Although now widely used, correlation coefficients had only been invented less than 20 years earlier by Sir Francis Galton, the cousin of Charles Darwin. If two variables go up and down together the coefficient has a value of 1, if they are totally unrelated the value is zero, and if they change in exact opposition the value is -1. As an example, the correlation coefficients between the plankton and the GSNW index in Figs 1.1, 1.3, and 1.4 are between 0.5 and 0.65.

Walker's initial attempts were hampered by the sheer volume of data that had to be processed by hand, but everything unexpectedly changed with the advent of World War I. When many of the department's senior staff were drafted off to Europe, and the few who remained were preoccupied with war-related matters, dozens of eager Indian civil servants normally assigned to other staff were free to work for him. Foremost among them was Hem Raj, who possessed a photographic memory for weather charts. After extensive sorting through world weather records and carrying out a mass of computations, Walker discovered three apparent 'seesaws' in atmospheric pressure and rainfall across the world's oceans, where a decrease in one region was matched by an opposite increase in another. His study of world weather, published in 1923, summarised this situation as:

There is a swaying of pressure on a big scale backwards and forwards between the Pacific Ocean and the Indian Ocean; and there are swayings, on a much smaller scale, between the Azores and Iceland and between the areas of high and low pressure in the North Pacific: further there is a marked tendency for the 'highs' of the last two swayings to be accentuated when the pressure in the Pacific is raised and that in the Indian Ocean lowered.[2]

Neither the first of these lesser swayings, that between the Azores and Iceland, nor the second, between the Hawaiian Islands and Alaska, held any clues to the mystery of the Indian monsoon collapses. However, the swaying between the Pacific and Indian Oceans looked more promising. Every few years the normal atmospheric conditions across the Pacific reversed, the usual high pressure in the east dropping at the same time as the low pressure in the west was raised. To distinguish it from the more northerly swayings, Walker called this one the *Southern Oscillation*.

Walker proposed that the normal atmospheric circulation pattern over the tropical Pacific is composed of a cell analogous to a Hadley cell, but orientated east–west along the equator. In this cell, surface winds converge on the low pressure generated by the warm sea and land surfaces of the west Pacific and Australasia, respectively. The converging air is forced to rise, leading to widespread cloud formation and rain. The Walker circulation is completed by a westerly return flow in the upper atmosphere and sinking of air over the eastern Pacific. Fluctuations of this circulation were then what caused the swaying of the Southern Oscillation.

This atmospheric cycle showed great promise for forecasting the variability of the monsoons—in particular, during the drought years 1877 and 1878, there was a strong pressure reversal over the equatorial Pacific. Subsequent observations have confirmed that, although weak monsoons occur in other years, pressure reversals in the Pacific are almost invariably associated with weak or low Indian rainfall. Unfortunately, without a physical mechanism to underpin this connection, further progress was not possible and it was not for another 50 years, at the very end of Sir Gilbert Walker's life, that an explanation began to be uncovered, beginning with the work of Norwegian–American meteorologist Jacob Bjerknes.[3]

After an early collaboration with his father Wilhelm, one of the founders of modern weather forecasting, Jacob had achieved scientific recognition for his work on the lifecycles of air masses and tropical storms. Half a century later, he turned his attention to the Southern Oscillation and how it might be linked to observations made by Peruvian fisherman. For centuries the anchoveta fisherman of Paita on Peru's northern coast had observed a warm current sweeping down from the north along the Pacific coast each year. In 1892, a Peruvian sea captain, Camilo Carrillo had published a short paper in the *Bulletin of the Lima Geographical Society* in which he reported how this warm current disrupted the rich anchovy fisheries close inshore: 'The Paita sailors, who frequently navigate along the coast in small craft…name this counter current the current of El Niño (the Child Jesus) because it has been observed to appear after Christmas.'

For a few weeks each year, this current would override the usual cool, north-flowing waters of the Humboldt Current. (Alexander von Humboldt, who made important contributions to many areas of science, sailed upon this current in 1803 and made measurements of its properties. Humboldt's book on the voyage was one of the books that Darwin made room for during his expedition on the *Beagle*. Ironically, although observing the current was only a modest portion of Humboldt's accomplishments, it may have become the thing for which he is now most famous.) The Humboldt Current is rich in nutrients because water is upwelling into it from depth. Every few years, the warm counter current extends much further south, and not for just a few weeks but for several months, bringing with it exceptionally heavy rains. Because the warm counter current is nutrient-poor and caps off the upward supply from below, there is much less plankton food for fish, and these events are barren times with dramatic reductions in fish catches. The lack of fish causes widespread starvation of seabirds.

In investigating this problem, Jacob Bjerknes had the advantage of access to a data set that at the time was unique. The year from 1957 to 1958 had been declared by the United Nations as the International Geophysical Year, and this led to the funding of a series of monitoring observations in all the world's oceans. By chance, this year also suffered one of the strongest El Niños of the century, with the result that scientists were provided with data on what was happening far out into the Pacific during one of these events. Looking at these observations, Bjerknes noticed that, just as Peru was experiencing the warm sea temperatures and coastal flooding of an El Niño year, the trade winds across the Pacific were weakened. Crucially, at the same time Gilbert Walker's atmospheric circulation had flipped. Although Bjerknes proposed this connection on the basis of what was still an incomplete data set, subsequent research has confirmed his insight that the observations of the Peruvian fishermen and those of Gilbert Walker in India are different components of a single large-scale phenomenon, a process which has subsequently been called the El Niño Southern Oscillation, or ENSO for short.

I experienced first-hand the fringes of a severe El Niño during the winter of 1997–8. In January, I arrived with my wife in San Diego to attend a Symposium and parked our hire car in the hotel car park. Sleep did not come easily after the trans-Atlantic journey, and so I wasn't pleased to be awoken in the early hours by the hotel management. Outside it was raining heavily and the receptionist told us our hire car was about to be flooded; the little river which had been inconspicuous behind the hotel upon our arrival, had become a swollen torrent. Some cars were already under water. This exceptionally extreme flooding was very unusual, and was attributed to the El Niño that was in progress. Further inland, in Arizona, the unseasonable rainfall became intense falls of snow. While I was getting wet moving the car, across the border in Mexico, at least a dozen people were dying in a mud slide. These were unfortunately only a few of the estimated 2100 people killed by weather events in the winter of 1997–1998.[4]

It is now known that El Niño, and its opposite state La Niña, are caused by irregular oscillations of the coupled ocean–atmosphere system across the whole of the equatorial Pacific Ocean. The eminent US oceanographer George Philander has called El Niño a dance between the atmosphere and the ocean. He has also expressed the coupling in symphonic terms: 'Acting in concert, the ocean and the atmosphere are capable of music that neither can produce on its own. El Niño is an example of such music.'[5]

The melodic ballet is a perturbation of the usual climatic pattern across the equatorial Pacific Ocean. The trade winds continuously push surface water westward, sustaining a pool of warm water thousands of kilometres across. To the east near South America, the warm water is replaced by cold water which flows from the depths to the surface, and in consequence the eastern Pacific is cold, even close inshore. This difference between the two sides of the ocean is reflected in the weather patterns. Little moisture evaporates from the cold ocean and so rain clouds rarely form. The Peruvian coast receives almost no rain and California and Mexico's Baja Peninsula have long dry seasons. Over in the western Pacific, heat and humidity can be uncomfortable and the moist air heated by the warm ocean forms massive rain clouds. These conditions continue until the monsoon bursts with a deluge of rain over South East Asia and Indonesia, bringing the waters needed for agriculture.

For some reason, every few years (usually in the southern hemisphere's spring) the trade winds slacken and sometimes even stop completely. As the winds abate the warm water piled up by the trade winds in the western Pacific flows backward to the east. It flows over the top of the cooler water and the sea surface is dramatically warmed. This is an El Niño. Rain clouds form over the Peruvian coast and these can be followed by torrential rain. Meanwhile surface temperatures cool in the western Pacific, inhibiting rainfall in South East Asia and Australia. The vast expanse of warm, moist air over South America disrupts the air flows around the Earth. Heavy rains and storms are brought to much of the North American west coast (as I experienced in 1998). North-eastern Brazil and the southern margins of the Sahara can suffer droughts. As the El Niño begins to reach global proportions, it affects the Asian monsoon, and the rains that fall throughout India and Pakistan from June to September may fail to appear, as happened so tragically in 1877.

No one knows what triggers an El Niño event. Changes in the Indian Ocean, in the snow cover in Asia or in Antarctic pack ice, seismic activity and volcanic eruptions have all been suggested as possible initiators. But the trigger may be less remote. It could be the changes in sea surface temperatures caused by the passing of a sub-surface wave, still reverberating from a previous event; or it could be one of the sudden bursts of westerly winds that interrupt the trade winds from time to time before disappearing just as suddenly. When Tony Busalacchi and colleagues at the University of Maryland looked at salinity data from the 1980s and 1990s, they found that low salinity levels in the western Pacific were followed by El Niño conditions six months later and excess salinity north and south of the equator with warmer waters about 12 months later. It could be that the first thing to trigger an El Niño event is the sinking of cold, salty water on either side of the equator which is then followed by the spreading of warm, less salty waters eastward.[6] Whatever the specific trigger, the basic cause is that the atmosphere and ocean in the tropical Pacific are never quite in equilibrium, and may sometimes be particularly sensitive to disturbance. At such times, even a gentle prod may be enough to generate a violent response from the wild beast.

Changes at the ocean surface caused by the faltering of the trade winds are spread across the ocean by two kinds of waves propagating within a few tens of metres of the sea surface. The first of these are Kelvin waves, which travel eastwards close to the equator and, rather like a conveyor belt, transport parcels of warm water towards the South American coast. Kelvin waves are named after the physicist Lord Kelvin who first described them. Kelvin's major achievement was establishing that heat is a form of motion. Unfortunately, he tends to be remembered even more for saying in 1900 that there were only two clouds remaining over the theory of heat and light:[7] one of his blemishes subsequently led to the theory of relativity and the other to the quantum theory.

The Kelvin waves further depress the thermocline in the east. In the open ocean the Earth's rotation only allows Kelvin waves to move along the equator. Here they travel at 100 km/day and can take up to two or three months to reach the coast. Once there, they can move north and south along the coast, distributing heat to mid latitudes. The equatorial Kelvin waves are partially reflected from the coast back across the Pacific as a second kind of sub-surface waves, Rossby waves. These waves are the oceanic equivalent of the atmospheric waves described in the last chapter. They travel westward in mid latitudes at only one-third the speed of Kelvin waves, taking close to 12 months to cross the ocean.

Upon striking the western coasts these waves are in turn reflected back as more Kelvin waves, but of opposite amplitude to the initial waves. It is commonly thought that these reflected waves provide the negative feedback which brings an El Niño event to an end. If so, it is the length of time it takes for these undersea waves to traverse the breadth of the Pacific basin that determines the year-long duration of a typical El Niño event. However, the El Niño cycle is a combined oscillation of the atmosphere and the ocean, and these changes in the ocean also have an effect on the trade winds.

At the end of the El Niño, the trade winds strengthen once more and warm water is pushed further west, restoring the normal balance. After most—but by no means all—El Niños the feedback system overshoots leading to the opposite extreme of the Southern Oscillation, El Niño's little sister, La Niña, 'the girl-child'. When this happens, all the typical features of the normal Pacific are exaggerated. The trade winds are stronger than usual, the pronounced low pressure over Australasia brings heavier rainfall to the region, and there is even more vigorous upwelling from the Peruvian depths. At its peak the warm water pool in the western Pacific can stretch as far as one-third of the globe and the extra heat loss can raise the global temperature by over 0.3°C. It has even been suggested that an El Niño event serves as a release valve for the heat build-up in the warm pool, while the La Niña event acts to recharge it.

There are written records of El Niño's effects in Peru at least as far back as the time of the Spanish conquistadors in the early 16th century, and there is geological evidence of El Niños in Peruvian coastal communities from thousands of years earlier. The Inca knew about them, and so built their cities on the tops of hills and kept stores of food in the mountains. When they did build on the coast they avoided the rivers. This is why so many of their dwellings are standing today. But it was not until the last quarter century or so that the rest of the world has come to realise that the phenomenon which involves only one-fifth of the circumference of the planet transforms the tropical weather all around the globe.

Perhaps the most important species affected by El Niños is the anchoveta, *Engraulis ringens*. In the Peruvian upwelling system, these fish mainly feed directly on the phytoplankton, which are at the bottom of the food web, and in turn provide the main food for a diverse array of predators at the top, including other fish, mammals, and large populations of oceanic birds. The fish's average lifespan is only three years but the females make the most of this by reaching spawning condition exceptionally early, at less than one year. On average, 50,000 eggs are laid over a female's lifetime. Unlike many other fish species, the anchoveta egg contains very little yolk, and so the anchoveta larvae are very dependent on a rich and immediate source of food. The fish is finely adapted to exploit the zone of high productivity, which is in the thin band of the northward flowing coastal current. The drawback of this adaptation is that no other fish is more vulnerable to the current's disappearance.

Many local seabirds are devastated when the anchoveta disappear: the Peruvian booby *Sula variegate*, the guanay cormorant *Phalacrocorax bougainvillii* and the Peruvian pelican *Pelecanus thagus* abandon their nesting grounds and leave the area, most flying south to Chile. In the El Niño of 1997, a Japanese film crew arrived on Chincha Island expecting to find 800,000 nesting birds, as they had on a previous visit, but on this occasion they could locate only a thousand. In such times, corpses of seabirds, along with those of mammals, fish and invertebrates, litter beaches in such quantities that they become a health hazard. South American sea lions and fur seals die because the squid and small fish on which they feed swim deep in search of cooler waters, and are then out of the animals' diving range. All this mortality extends as far north as California. The high concentrations of hydrogen sulphide given off by decaying bodies may blacken ships' hulls, a phenomenon called locally *El Pintor*, the Painter.[3,4,5] A colleague informed me that, in one shipping line, it was a disciplinary offence to waste white paint by using it in El Niño conditions.

The 1997–8 El Niño, perhaps the climatic event of the century, illustrates just how dramatic and far-flung the impacts of a strong El Niño can be. According to the Pacific Marine Environment Laboratory in Seattle every month between June and December 1997 set a new monthly record high for sea surface temperatures in the eastern equatorial Pacific. In Chile, rain fell on parts of the Atacama desert in amounts not seen since the arrival of the Spanish. Texas suffered extreme heat, and Laguna Beach in California experienced weather it had never previously encountered. There was snow in Mexico and an 'ice storm' in Canada that cut off much of Montreal's electricity for a week. It was also blamed for the worst fires on record in Indonesia, fires that were held to be responsible for a dramatic increase in respiratory disease in parts of Malaysia; Sudan and the Philippines suffered droughts that caused famines.

El Niños typically cause a decline in the palm oil production of the Philippines. In Africa, the altered wind, heat, and moisture patterns of El Niños commonly portend drought, generally in the east and south, a shift partly due to cooling of the south-western Indian Ocean strengthening the high-pressure area that keeps rainfall from reaching the south. Unusually, in 1997–8 much of East Africa was drenched.

With ENSO cycles, the negative effects of one phase of the cycle will be balanced over time by the positive effects of the other. As a consequence, those species with the greatest ability to conserve resources, switch prey, or travel to refuges elsewhere will find it easiest to ride out negative effects without a pronounced impact on their populations. The most dramatic impacts will be experienced by species tied to a location through breeding requirements (seabirds), dietary specialisation (anchoveta), or lack of mobility. Inability to move away is the reason that one of the most severe and widespread biological impacts of El Niños such as that in 1997–8 is often on coral reefs, with corals once vibrant with kaleidoscopic colour becoming bleached white as if blanketed in a winter's snow. The colour of the corals comes from the pigmentation of zooxanthellae, microscopic algae that live symbiotically within the coral tissue. These algae provide food for the coral, which in turn acts as a host. Bleaching is caused by the eviction of pigmented algae, exposing the white calciferous skeleton through the transparent coral tissue.

Relatively small rises or falls in temperature will cause most species of coral to expel their zooxanthellae. Corals can generally tolerate short bursts of stress, and as long as some reserves of zooxanthellae are retained, the coral can bounce back but, if the stress is severe or prolonged, all the algae are permanently evicted and the coral dies. In 1997–8, corals living in Australia's Great Barrier Reef expelled their algae, becoming white. Some areas recovered but along an 18-mile length in Western Australia the corals died. Abnormally warm waters caused 90% mortality of corals in the shallow zones of many reefs off Sri Lanka, India, the Maldives, Kenya, Tanzania, the Seychelles, and the Galapagos Islands, while at the same time abnormally low temperatures killed reefs in Indonesia.

Not all the impacts of El Niño spell disaster, however. In the late 1990s, beetles imported to Lake Victoria, which borders Kenya, Tanzania, and Uganda, were credited with clearing water hyacinth, an introduced plant that was choking the lake and harming the economies of the surrounding countries. It now seems that the beetles may have had help. Adrian Williams, working for the environmental consultancy APEM in Manchester, UK, has shown that the increased rain and reduction in light during the El Niño event in 1997 would have caused the plants to grow poorly and taller, making them more susceptible to the beetle's depredations. Without the El Niño the beetles might have been less successful and, following its end, the hyacinth is again beginning to spread over the lake.[8]

But, for the countries of Asia, El Niños sometimes do spell disaster. As Sir Gilbert Walker and Jacob Bjerknes discovered, the crop-watering rains of the Asian monsoon are prone to fail when El Niño warms the surface waters of the Pacific. This does not happen with all El Niños, however. It seems that the manner in which the ocean warms up is critical. Krishna Kumar at the Indian Institute of Tropical Meteorology and colleagues[9] have found that the risk of monsoon failure depends on where the waters warm most: raised temperatures in the central Pacific are more harmful than those farther east. This means that detailed measurements of sea surface temperatures in this region may improve forecasts of India's monsoon, something Gandhi could only hope for at the time he wrote his letter.

The Southern Oscillation in the Pacific Ocean provided the answer that Gilbert Walker was looking for when he examined patterns in the world's weather, seeking a way of predicting when the Indian monsoon would fail. However, he also came across two other global seesaws which proved not to be useful: the North Pacific Oscillation (NPO) and the North Atlantic Oscillation (NAO). These oscillations also significantly impact on living systems. The NPO, now commonly subsumed within the wider Pacific–North American (PNA) pattern, is associated with weather and biological changes over the extra-tropical North Pacific. In this pattern, warm sea surface temperatures over the Pacific west of 180°W influence the atmosphere to produce a wave-like set of anomalies which propagate across the western hemisphere. The effect is to produce warming of the continental United States, but a cooling of western Canada and the Canadian Arctic. There is an increase of cyclones and precipitation over the warmed regions at the expense of the Canadian Arctic. The reverse chain of processes also occurs. However, these climatic changes over the North Pacific are not as regular as those over the North Atlantic, because the episodic ENSO events significantly disturb the climate of the Pacific basin, and so the PNA is not a separate oscillation to such an extent as the NAO or the SO.

The NAO is a feature of the winter weather patterns over the North Atlantic, and its strength is measured by the NAO index calculated in this season. The pattern is not really seen in data from the summer. Even

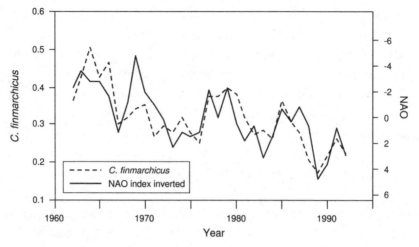

FIG 6.1 Numbers of *Calanus finmarchicus* in the NE Atlantic and North Sea, 1962–1995, (solid line) compared with the NAO index (broken line). (Redrawn from Fromentin and Planque, 1996.[10])

so, its effects are widespread in weather, and plant and animal populations over the North Atlantic region and across Europe, and they are seen in data from the CPR Programme. Jean-Marc Fromentin and Benjamin Planque from Villefranche-sur-mer, France, have shown that the abundance of one particular copepod, *Calanus finmarchicus*, in the NE Atlantic and North Sea seems to be dependent on the strength of the NAO.[10] This animal is an important food for young stages of commercial fish such as cod. As Fig. 6.1 shows, a strong NAO was associated with low numbers of *Calanus* and weaker values with greater numbers.

The NAO is essentially a winter phenomenon and *C. finmarchicus* has a unique lifestyle among the copepods (already described in Chapter 3) which makes it especially susceptible to it. Rather than struggling through each winter in the surface waters of the sea, it migrates to cold water at several hundred metres down in the North Atlantic to hibernate through the winter, only returning again in the spring. The copepods then have to move back into the North Sea around the north of Scotland at the start of the year, a journey which is dependent on the winds at

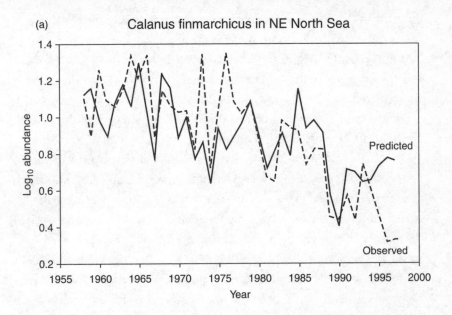

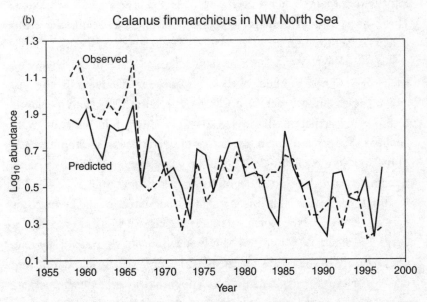

FIG 6.2 *Calanus finmarchicus* numbers observed in the NE and NW North Sea (solid lines) compared with predictions based on current flows.[11]

this time. As a consequence, numbers of the copepods in the North Sea can be predicted from strength of the current flows into the North Sea during the winter (Fig. 6.2).[11] The relationship between C. finmarchicus and the strength of the NAO arises mainly from this dependence on winter winds.

Other populations going up and down with the strength of the NAO also tend to be particularly susceptible to conditions in winter. Thus, when Martin Attrill and Michael Power observed that over a 16-year period the NAO was the most important factor explaining the composition, abundance, and growth of juvenile fish in the Thames estuary, they attributed this to the fish exploiting the temperature difference between the estuarine and marine waters during the winter.[12] Fish preferred to spend this season in the Thames, because its water was cold, and this reduced their metabolism and food requirements at a time when food was scarce. This response of fish to temperature is consistent with laboratory experiments, and has even been observed in an analysis of angling. This showed trout in a small reservoir moving to cooler deep water at the height of the summer.[13]

Events in the North Atlantic are also linked to those in the equatorial Pacific. ENSO cycles influence the path of the Gulf Stream. In the last chapter, it was shown that the position of the current is mostly a delayed response to fluctuations in the NAO (Fig. 5.1). Much of the remaining variability in the Stream's path is a similar delayed response to the strength of the southern oscillation.[14] About two years after an El Niño, the Gulf Stream tends to be shifted slightly northward, and after a La Niña it is shifted southward. But this is not the total extent of the connection between ENSO and the North Atlantic. Anastasios Tsonis and colleagues at the University of Wisconsin-Milwaukee have shown statistically that the variations of the ENSO and the NAO become synchronised for a few decades before the links abruptly break down and a new pattern emerges. They called it 'synchronised chaos'. It seems from their modelling studies that these actions are always driven from the North Atlantic,[15] and this raises concerns if a climate tipping point resides in these waters. An abrupt change in the North Atlantic could then trigger some shift in the ENSO cycles with ramifications around the globe.

All three of the oscillations discovered by Gilbert Walker involve some degree of coupling between the atmosphere and the ocean, and many of these processes in the ENSO cycles have now been worked out. Further north in the Pacific and Atlantic Oceans, modulation of air–sea heat transfer also allows the possibility of cycling from one state of the ocean and atmosphere to another. The surface of the ocean may warm over winter if heat loss from the ocean's surface is reduced. Eventually, the warm ocean will pump heat into the atmosphere, initiating cyclones and strengthening the surface winds. These will then take heat out of the ocean, eventually cooling the surface temperatures and pushing the climatic state towards a cold extreme. In this way, cycling back and forth, such as in the NAO, may arise. Ocean currents are certainly forced by the atmosphere, but do the oceans this far north influence the atmosphere in turn?

The question of what impact the state of the ocean in mid latitudes has on the atmosphere is also raised by the linkage between the latitude of the Gulf Stream, set up by years of weather, and the plankton populations thousands of miles further east whose lives are completed in months. One possibility is that the ocean currents in the North Atlantic modify the weather patterns over it in some way to which the plankton respond. But how much are the air and sea intertwined this far away from the equator? The search for mutual dancing of the atmosphere and oceans in more northerly latitudes begins in a place where the sun behaves strangely.

7

ACTION AT A DISTANCE

The cause of remarkably mild winters which occasionally occur in England

'the extension of the Gulf Stream in that year to the coast of Europe, instead of terminating about the meridian of the Azores. In the winter of 1821–2, the warm water of the Gulf Stream spread itself beyond its usual bounds over a space of ocean which may be roughly estimated as exceeding 600 miles in latitude and 1000 in longitude, carrying with it water several degrees higher than the temperature of the sea in ordinary years in the same parallels.'

(Lieut.-Colonel Sabine, 1846)[1]

Novaya Zemlya is the land where the sun is rectangular. This archipelago of islands in the Arctic Ocean, separates the Barents Sea (the Arctic Ocean north of Norway and Finland) from the Kara Sea (the Arctic Ocean north of Russia). They are part of Russia, and their Russian name, mainly applied to the larger of the two main islands, means *new land*. The northern island, which is blanketed by glaciers, covers about 50,000 square km, while most of the 40,000 square km of the southern island is a treeless plain. Both islands have large deposits of coal, together with some copper, lead, and zinc.

Early Russians first discovered Novaya Zemlya about 1100 AD, but the islands remained uninhabited until 1877, when an Arctic people called Samoyeds established the first permanent settlements. Nowadays, a small colony of Russians and Samoyeds live on the southern island, raising reindeer, trapping and hunting animals, and collecting eider down (the feathers of eider ducks). The northern island is less hospitable because it still shows harmful levels of radiation from nuclear tests carried out by the Soviet Union during the 1950s.

In 1597, the Dutch explorer Gerrit de Veer was the ship's carpenter with William Barents' (or Willem Barentsz) third expedition into the Arctic. Barents, a Dutchman for whom the Barents Sea is named, is one of the more renowned of the early European explorers of the Arctic. Around the end of the 16th century, several expeditions tried to establish a north-east passage to Asia, but fog, pack ice, and the confusing geography of the coast of Russia frustrated all efforts. No one pursued this search with such tenacious skill as Barents, who embarked on three separate expeditions along the frigid and unforgiving northern shores of Eurasia seeking the illusive passage. On his first two voyages, Barents reached the archipelago of Novaya Zemlya, on the way rediscovering Bear Island and Svalbard (midway between the northern tip of Norway and the North Pole) before sailing east into the Kara Sea. His third attempt, in 1596, was disastrous.

During this expedition, his ship became trapped by sea ice, and he and his crew of 16 men were forced to winter ashore on Novaya Zemlya, building a crude cabin from the wrecked ship. Battling scurvy, hunger, the assaults of bears, and the intense winter cold, they were the first west Europeans to winter in the high Arctic and survive. Then, in June 1597 when the ship was still not released by the ice, Barent's party set out on a 1600-mile escape in two open boats. Barents died on this journey, but many of his men survived. Among the survivors was Gerrit de Veer, and he went on to chronicle the voyages in great detail, including in the account a unique observation.[2]

In his diary, de Veer described how, as the long polar night draws to a close, the sun will suddenly burst above the horizon weeks ahead of its astronomical schedule. This is basically a polar mirage in which a strong, shallow, surface-based atmospheric inversion layer, acting as a mirror, reflects the light of the sun when it is just below the horizon. It has since become known as the Novaya Zemlya effect. The mirage occurs when the lowest layer of air is colder than the layer lying over it, an arrangement which refracts rays of light downwards. For it to happen, the rays of sunlight must travel through the inversion layer over hundreds of kilometres. To elevate the sun's disc by the observed amount—about 5 degrees—the sunlight must follow the Earth's curvature over at least 400 km, a distance that is determined by the temperature gradient in the inversion layer. Depending on the meteorological conditions, the effect will present the sun as a line or a square (the 'rectangular sun'), sometimes made up of flattened hourglasses.[3] Gerrit de Veer was the first person to describe this phenomenon.

William Barents' winter camp on the northernmost tip of Novaya Zemlya remained undisturbed for almost 300 years. Then, in 1871, the house in which Barents and his crew wintered was discovered, along with many relics (now preserved in The Hague), by the Scandinavian explorer Elling Karlsen. At this time more and more boats were competing for prey such as walrus, whose hides were in great demand for making drive belts for machinery, and this had led to a drastic decline in numbers. In a search for new hunting grounds, Elling Karlsen sailed eastwards during 1868, eventually discovering rich grounds around Novaya Zemlya. During the following three seasons the crews of the Norwegian ships made a series of observations which greatly added to the accumulated knowledge of the surrounding waters, recording water temperature and depth, ice conditions, and the general character and geographical position of the coastlines. In the course of doing this, they made an unusual discovery.

On a group of islands off the northern part of the archipelago, they found wooden floats from Norwegian fishing nets, beans from the West Indies, pumice stone from Iceland and the wreckage of ships. The pumice stone must have been bobbing up and down in the current for 3000 km, and the beans could have been doing so for three times as far. This was evidence that the influence of the Gulf Stream and its northward extensions can be registered as far away as north of the middle of Russia and as close to the pole as almost 80°N. The islands subsequently became known as the Gulf Stream Islands. It is even possible that these bits of flotsam and jetsam are accompanied by heat. Gier Ottersen, Bjørn Ådlandsvik, and Harald Loeng from the Institute of Marine Research in Norway wished to forecast temperatures in the Barents Sea to help with the management of the fisheries in the region. Having unsuccessfully tried other predictors, they discovered that, once they had made allowance for the persistence of temperatures from one year to the next, half of the total historical temperature variability could be explained by a combination of the atmospherically driven flow into the western Barents Sea during the preceding year and the position of the Gulf Stream (as measured by the GSNW index) two years earlier. These time lags are consistent with the time taken for patches of warm or cold water to drift with the ocean currents over the respective distances.[4]

The objects found on the Gulf Stream Islands were transported by an extension of the same warm currents that transport organisms such as 'blue snails' and the Portuguese man-of-war from the tropical regions to the shores of the British Isles. Such transport by the currents of the Gulf Stream system plays a vital role in the lives of some animals. Eels start their life in the Sargasso Sea before migrating into streams, rivers and lakes around the ocean. At this stage they are less than 90 mm long and are called glass eels. Some go to the US coast, but others drift with the Gulf Stream and the North Atlantic current to the freshwaters of the British Isles and Europe. They then spend the next 10 years growing in the rivers and lakes, before the silver eels (as they have become) begin the return journey to breed in the Sargasso Sea. Most of those from the USA make it, but it is not known how many from Europe successfully navigate the 6000 km of currents that are flowing against them. By the time they leave the continent, each eel's gut has dissolved, making feeding impossible, and so the European eels have to rely on stored energy alone for the 150 or so days their journey is likely to take.

Another animal migrating with the same currents is the leatherback turtle, the largest of the turtles and an animal that is also found in the Pacific and Indian oceans. It is unique in having a leathery carapace, rather than a shell, and in being warm-blooded. The tiny turtles emerge

from a nest one metre down on a sandy beach in the subtropics or tropics. Small and vulnerable, the scaly, 80-mm hatchlings have to reach the
sea quickly, for a range of predators await. If they survive this journey,
the turtles can grow to over 2 m long, feeding mainly on the jellyfish that
drift with the ocean currents. Initially they live in the Sargasso Sea, but
some follow the Gulf Stream system to the Bay of Biscay and the seas
off southern Ireland. It is known that they appear amongst the ice floes
off Newfoundland, and they have been recorded off Iceland and Norway.
Eventually, they travel along the coast of Portugal and Africa, returning
to the Caribbean to breed.[5]

But the most important effects of the currents are commonly thought
to be climatic. Only about 700 km west of Novaya Zemlya, the ocean is
permanently ice free, even though it is at 75°N, and north of Norway and
Russia. This is a long way north of areas in Canada that are heavily frozen
each winter. The climatic amelioration is mainly due to the transport of
heat by winds warmed by the ocean, but also to the currents that continue from the Gulf Stream to the north of Scandinavia. The Gulf Stream
has been shown to influence the atmosphere above it by causing conver-
gence of surface winds and thereby anchoring a band of precipitation
along it.[6] In this rain band, upward motion and cloud formation extend
to several kilometres in height with the frequent occurrence of very low
cloud-top temperatures. These mechanisms provide a pathway by which
the Gulf Stream can affect the atmosphere locally, and possibly in remote
regions by forcing planetary waves.

The first account of the climatic influence of the Gulf Stream was published by the American naval lieutenant, Matthew Fontaine Maury, in
1856. He wrote that: 'One of the benign offices of the Gulf Stream is to
convey heat from the Gulf of Mexico, where otherwise it would become
excessive, and to disperse it in regions beyond the Atlantic for the amelioration of the climates of the British Isles and all western Europe', and
he said that, without the Gulf Stream, 'the soft climates of both France
and England would be as that of Labrador, severe in the extreme and ice-
bound'. He was not alone, for around the same time Robert FitzRoy also

wrote that Britain's climate very much depends on the Gulf Stream and the warm, moist westerlies that it brings with it. He also added somewhat prophetically that, should this 'peculiar circumstance' alter, because of some change such as a 'diminution of ice in the polar regions', there is a 'fear of a gradual change in our average climate'.[7] In the early 1900s there were even speculative proposals (for example by the American, C.L. Riker, and the German, J.E. Kiesel) for climate modification by manipulating the flow of the Stream.[8]

It was natural to suppose that, if the flow of the Gulf Stream was making Europe's climate mild, the currents and temperatures of the North Atlantic may have influences on the continent's weather. In an 1846 paper 'On the cause of remarkably mild winters which occasionally occur in England', Sir Edward Sabine, an Irish scientist perhaps best known for his work on the Earth's magnetic field, was one of the first people to speculate on the possible effect of anomalous strength of the Gulf Stream upon temperatures prevailing in the following winter in Europe.[1] This was picked up 70 years later by P.H. Galle in Holland who found that anomalies in the trade winds in the North Atlantic during the years 1899 to 1914 were correlated with air temperatures over Europe some time later, and speculated that changes in the Gulf Stream provided the connection.

Although generations of schoolchildren have been told about the warming effects of the Gulf Stream ever since the days of Maury and FitzRoy, the actual situation is more complex, so much so, that Richard Seager from the Lamont Doherty Earth Observatory at Columbia University in New York has described it as a myth, the climatological equivalent of an urban legend. Seager has pointed out that both observations and climate models show that the atmosphere does the lion's share of transporting heat northward to the latitude of Europe. When he and his colleagues used a general atmospheric circulation model to quantify the climatic role of the Gulf Stream, they found that the Stream accounted for no more than 10% of the temperature difference in winter between Britain and Newfoundland.[9] The remaining 90% was made up of two contributions: the maritime influence of the relatively warm body of surrounding water, an influence which has nothing to do with the currents of the Gulf Stream system, and the bolstering of this maritime effect by south-westerly winds bringing a warm air mass from the south.

Water retains summer heat far longer than land, which is why the winter–summer difference in temperature is about 5°C over the North Atlantic and yet nearer 50°C at the same latitude in Siberia. The storage and release of heat in different seasons accounts for almost half the winter temperature difference across the North Atlantic Ocean. Even without the Gulf Stream, western Europe would be in the warmth brought in by the prevailing westerly winds from the heat stored in the Atlantic Ocean. This warming effect is not unique to the Atlantic Ocean for a very similar process occurs across the North Pacific Ocean. In the Atlantic, the warming is amplified because the winds are not due-westerlies but come from the south-west and so transfer additional heat from the south.

The origin of these south-westerlies can be traced to a massive north–south meander in the wind patterns over North America, a meander generated by the Rocky Mountains. When Seager and colleagues removed the Rockies from their computer models, making the Western USA flat, the temperature difference in winter between Newfoundland and Britain was reduced by about 9°C. Because of the need to conserve vorticity, as air flows from the west across the mountains, it is forced to the south and then to turn back to the north further downstream, thereby following a Rossby wave. In this way, the mountains force cold air south into eastern North America and warm air north into western Europe.

While the Rocky Mountains may be a major cause of the mild temperatures in western Europe they do not generate the fluctuating weather patterns experienced in the region, for the height, position, and shape of the mountains only vary on geological timescales. Further, even though the effect of the Gulf Stream itself may be small, its combined effect with the heat storage of the ocean still accounts for half of the European warming. Consequently, even if the weather of western Europe may not be impacted by the Gulf Stream to the extent envisioned by Sabine and Galle in the 19th century, the oceans still have a significant influence. If so, perhaps the weather patterns to which the plankton and other populations are responding might be caused by changes in the ocean that are happening at the same time as the movements of the Gulf Stream.

But could the oceans be the source of many weather events? For unusual conditions in the ocean to noticeably influence weather patterns, they must be of a certain scale and intensity. Forty years ago, the UK meteorologist J.S. Sawyer summarised this general requirement more specifically in three criteria.[10] The anomaly at the ocean's surface must: cover a region of 1000 km or more across, be strong enough to raise or lower the total heat input to the atmosphere by at least a tenth of the solar energy absorbed by the Earth (about 24 W m^{-2}), and persist for at least a month. Large features satisfying these conditions do occur in both the North Atlantic and North Pacific Oceans and so there is the possibility of the oceans impacting on the atmosphere. However, Sawyer's criteria are merely the minimum conditions for some impact; they do not guarantee one.

It was not until the second half of the 20th century that the possibility that the status of the ocean might be at the root of weather patterns downstream was explored in detail. While there were a number of other investigators involved, the name of Jerome (Jerry) Namias stands out. Namias was one of the world's greatest long-range weather forecasters, engaged on what he liked to call the world's second most complex problem. (He considered that predicting human behaviour was the most complex problem, and recent economic events tend to support his view.) A colleague has described him as 'A man who gives good reasons for any long-range forecast, and even better reasons for why it fails...a man who is an infinite source of good ideas...who thinks fast on his feet,...is always a scholar...and a gentleman.' While lacking formal meteorological training, Namias eventually received the highest award of the American Meteorological Society and established the long-range forecasting branch of the US Weather Service, and the Experimental Climate Prediction Center at the Scripps Institution of Oceanography. He may also have been the only person elected to the prestigious US National Academy of Science without an undergraduate degree.[11]

A major turning point for Namias' ideas occurred at the 1957 California Cooperative Fisheries Organization Conference in Santa Fe. This conference had been assembled because a remarkable oceanic warming (which we now know to have been an El Niño) had occurred over the eastern Pacific. The sea surface temperature abnormality was accompanied by disturbances in the California current, marine chemical properties, and also the marine biota, with southern fish being caught in northern waters. There were unusual typhoons and, in general, both atmosphere and ocean were behaving in a manner that was far from normal. Invited to be a principal speaker at the conference, Namias gave a talk diagnosing the unusual weather events in mid latitudes and then sat back to listen to the other speakers describing the subtle interplay of atmosphere and ocean in the tropical Pacific region. He later wrote, 'The inter-associations quickly became clear, and it struck me that some of the secrets of long-range weather forecasting might lie in the coupled air–sea system. It was especially noteworthy that the mismatch of timescales in the two media, air and sea, could account for the frequently observed long-term memory required for long-range problems.'

Following this meeting, Namias began increasingly to draw upon the influence of the ocean surface in his climatic studies. However, it was several years before he could actually begin to work full time on the questions of large-scale air–sea interaction and the effects of persistent temperature patterns in the ocean, efforts that were to occupy the rest of his life. The delay was partly because he had a coronary heart attack in 1963 and also because, even though he never actually learned how to drive, he was involved in a bad automobile accident in the following summer. He was also hindered by the difficulty of getting funding, for the developing field of numerical weather prediction began to dominate research in weather forecasting.

From a number of carefully analysed case studies, it became evident to Namias, who was now at Scripps Institution of Oceanography in

California, a laboratory looking across a sandy beach to the rollers of the Pacific Ocean, that if abnormally warm water was generated at high latitudes in the Pacific Ocean during the summer, the Aleutian low (the Pacific's equivalent of the Iceland low) would be intense in the subsequent autumn. Similarly, cold water in the summer would lead to abnormally high pressure in autumn. He reasoned that the increased heat flow from a warm sea surface would amplify cyclones by destabilizing cool air masses, whereas the opposite situation would occur over a colder than normal ocean. Namias further suggested that pools of anomalous water might be hidden at depths below the surface thermocline during the warm season. With the onset of increased storminess during the autumn, these subsurface anomalies could be brought back to the surface by vertical mixing and could then impact on weather patterns.

Namias also proposed that blocking in the North Atlantic, the occasional tendency for an anticyclone to persist off the coast of Europe or Scandinavia deflecting cyclones away, could be associated with unusual temperatures in the middle of the ocean. In 1970, this work was extended by R.A.S. Ratcliffe and R. Murray at the UK Meteorological Office who, from an analysis of many observations, showed that when water over the Grand Banks of Newfoundland is cold a high pressure develops in the atmosphere to the north-west of Britain giving warm weather, while a warm patch of water has the opposite effect.[12] Sea temperatures in this critical area were supposed to be dependent on the interaction of the Gulf Stream with the cold Labrador Current.

Then, in the late 1970s, there was somewhat of a backlash against all these ideas when careful statistical analyses by several groups, including some who were also at Scripps, indicated that the North Pacific Ocean might be no more than a passive responder to atmospheric forcing. These studies showed that it was much easier to produce anomalous sea surface temperatures (SSTs) from atmospheric anomalies than it was to produce atmospheric anomalies from SST anomalies. In addition, a number of attempts have been made to replicate Namias's and Ratcliffe and Murray's observations using computer models of the atmosphere.

These experiments have to be carried out with care. Each model run must be repeated several times, starting with slightly different conditions, so as to determine whether any signal is any more than the natural variability of the atmosphere. While there have been some successes in reproducing the observations,[13] other attempts have been less successful,[14] and overall the results have not been conclusive. Although general circulation models of the atmosphere showed that they were sensitive to what the SSTs in the tropics were, they did not seem to respond much to SST anomalies further north. It is possible that some of this insensitivity may have been due to the early generation of general circulation models (more recent models have since shown greater sensitivity to global SSTs) but it is still unlikely that downstream effects of abnormal ocean SSTs are strong.

Even so, the ocean is not always just a slave to the atmosphere; sometimes it does have some say in what ensues. Ocean temperatures have an effect on wildfires in the wilderness of the western USA. Because warmer oceans encourage warmer weather, temperatures in the equatorial Pacific have long been monitored to forecast the danger of wildfires in the south-western USA. But warmth in the Pacific Ocean may spark fires in the north-west as well. When Yongqiang Liu of the US Department of Agriculture's Forest Service in Athens, Georgia, incorporated the anticipated 0.6°C warming of northern Pacific by 2080 into two forecast models, the models predicted a sharp increase in the extent of fires.[15] The calculations showed that the land burned annually in the north-west could grow from under a million hectares in 2002 to nearly 2 million hectares by 2080. This happens because warmer than average ocean temperatures in the North Pacific create more low pressure weather systems than cooler waters do (as Namias had suggested), pushing the jet stream further north into Canada and allowing high pressure systems to move in from the south, bringing drier and hotter air to the north-west.

This is not the only such maritime effect in the USA. Song Feng's group at the University of Nebraska found in model simulations that a cooling of the Pacific by 3°C could cause a winter drought such as happened during the Dust Bowl of the 1930s by reducing the occurrence of winter storms. Meanwhile, a warming of 1°C in the Atlantic reduced transport of moisture in summer to the Great Plains of central USA and further to the west.[16]

All of the above observational and modelling investigations were concerned with relating weather patterns with conditions in the parts of ocean from which the weather systems have come, generally in a similar latitude range. While it is possible on some occasions to associate atmospheric patterns with the distant situation in the ocean, in general it appears that such associations are weak. In the case of the wildfires, it was their upward trend that was related to a steady warming in the ocean, while the US drought was a response to a prolonged period with ocean temperatures that were out of the ordinary. But things are very different in the tropics, especially in the Pacific. It is the two-way coupling of ocean and atmosphere that is central to ENSO cycles and, as a result, the state of the Pacific Ocean affects weather all around the equator and influences the Asian monsoons. Here the east–west distribution of heat across the ocean reinforces the wind pattern responsible for it until some trigger in the system initiates a reversal. Even so, like the ocean forcing of wildfires, the timescale of the coupling is months to years, and this also seems to be the case in the North Atlantic Oscillation.

Although primarily a winter phenomenon, the NAO is in many ways the cousin of ENSO. Fluctuations of the NAO are intimately related to the position and strength of North Atlantic storm tracks and consequently correlate with all sorts of surrounding observations, from rainfall in Bordeaux to the amount of Saharan dust that reaches the Bahamas and the abundance in the fisheries off Iceland. The status of the NAO affects the currents at the North Atlantic's margins and cyclones right across the northern hemisphere. However, computer models reveal that there are differences between the NAO and its cousin half a world away. When computer models of the atmosphere are run with temperatures at the sea surface held fixed, the weather patterns over the North Atlantic still show a flip-flopping very similar to the NAO. However, ENSO shuts down completely in such models because oceanic changes are an essential component of its mechanism. Therefore, at first sight the NAO appears to be an entirely atmospheric phenomenon.[17]

But there is more to the NAO than this. The NAO doesn't just fluctuate on timescales of weeks and months as tends to happen in such model runs; it also has slower rhythms of years, decades or longer. As described previously, the strength of the NAO is measured by an index, the difference in atmospheric pressure during the winter between the Azores and Iceland (the pressure features shown in Fig. 5.2). During the 1940s there were a series of low values of the index which accompanied some of the coldest European winters of the century, including those that delayed Hitler's invasion of France and helped defeat his assault on Moscow. In contrast, the strong NAO states in the late 1980s and early 1990s corresponded to a sequence of particularly mild winters across Europe. While decade-long runs of high and low values of the NAO do sometimes occur in simulations of the atmosphere alone, they are much more common in the observations. It seems likely that there is some long-term memory operating, and the natural first place to seek such a memory is in the ocean.[17]

When Jim Hurrell at the US National Centre for Atmospheric Research and Mike McCartney from the Woods Hole Oceanographic Institution in Massachusetts took a set of years with similar winter NAO strengths and looked at what the NAO was like in earlier and subsequent seasons, they found that there was no fixed pattern at all. In the warmer parts of the year, the difference in pressure between the Azores and Iceland appears to wander about haphazardly. But they discovered that when winter comes, the NAO tends to go back to the state it was in the previous year, no matter what it had been doing in the meantime. This connection between years could be provided by the ocean. Once the warm surface layers that developed during the summer are gone, the atmosphere is exposed to the ocean conditions remaining from the previous winter. The NAO may not be the same intricate dance of air and sea seen in ENSO, but its long-term shifts may be the product of a different complex dance with the same partners.

How the coupling of ocean and atmosphere operates in this system remains unclear. Certainly the NAO has pronounced effects on the North Atlantic Ocean. Cold winter winds blowing over the seas either side of the towering ice mass of Greenland, the Labrador Sea, and the Greenland Sea, cause the water to become cold and dense, generating cold salty water which spreads southward in the Atlantic at great depth. When Bob Dickson of the UK Centre for Environment, Fisheries and Aquaculture Science in Lowestoft examined observations from these two seas, he and his colleagues found a marked difference between what happened under weak NAOs in the 1960s and the consistently strong NAOs of the 1990s.[18] During winters with a strong NAO, westerly winds roaring down from the north across the Labrador Sea drive strong convective sinking. At this time, the winds over the Greenland Sea tend to be more southerly making the sea warmer and less salty, so that the water is less dense and sinking is reduced. When the NAO weakens, though, the tables are turned and there is a tendency for cold, dry winds to sweep down from the North Pole over the Greenland Sea. The same winds drive relatively fresh water and sea ice down the east coast of Greenland and round its tip into the Labrador Sea, where the freshness and lack of cooling make deep convection less likely. During the early 1990s, the strong NAOs drove convection in the Labrador Sea at a furious pace, producing large amounts of cool water at depth, while in the 1960s the weaker NAOs instead generated cold, deep water in the areas to the east of Greenland.

These oceanic changes are just part of a widespread pattern of shifts in the North Atlantic[19] but whether such changes in the ocean in turn react upon the atmosphere remains to be determined. It has been suggested that this cool water at depth could make its effects felt in the Gulf Stream passing over it. Alternatively, the Stream might be affected by variations in the cold Labrador current which the winds drive down the eastern coast of Canada from the North. In either case, the changes might then influence the atmosphere further afield in the manner described by Ratcliffe and Murray. Certainly, if computer models of the atmospheric circulation are supplied with data on how the ocean temperatures around the world have fluctuated, they are able to reproduce many of the past vagaries of the weather. A group led by N.S. Keenlyside of the Leibniz Institute of Marine Sciences in Germany has shown that using global ocean surface temperatures in this way improves the forecasting power over the next decade.[20] Their model predicts that natural climate variations in the North Atlantic will temporarily offset global warming, and Europe and North America may even cool a little over this period.

If atmosphere-only general circulation models are run using observed sea surface temperatures around the globe, they can reproduce the past trends in the NAO. Such calculations can then be used to probe the mechanisms in operation. These studies began in 1999 when a young researcher Mark Rodwell, in collaboration with colleagues Dan Rowell and Chris Folland at the UK Meteorological Office, used a general circulation model of the atmosphere to investigate the ocean's role in forcing North Atlantic and European climate.[21] The global atmospheric model used in the study was HadAM2b, a model with almost 7000 horizontal grid-boxes and 19 vertical layers. As with all such models, it takes account of the earth's topography, and it predicts both cloud cover and ice cover. When they ran the model to determine what atmospheric patterns were consistent with the worldwide changes in temperature at the sea surface and extent of sea ice observed from 1947 to 1997, they discovered that it reproduced the decline in the NAO which occurred to the 1960s, its subsequent rise to the early 1990s, and some of the subsequent fall (see the lower graph in Fig. 7.1). Moreover, the simulations captured some of the year-to-year changes in the NAO (the upper graph in Fig. 7.1). The model also reproduced much of the observed trend in winter European temperatures.

Therefore, low-frequency variability in the NAO is not merely stochastic atmospheric noise but is a response to oceanic conditions. But which region of the ocean is most responsible? Rodwell's group argued for it being the North Atlantic basin itself, very much along the lines proposed by Ratcliffe and Murray. To do so, they used their model calculations to show how North Atlantic sea surface temperature characteristics are

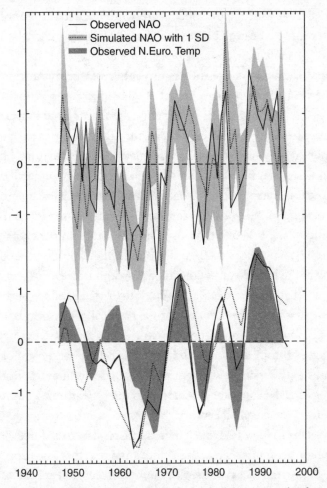

FIG 7.1 Rodwell, Rowell, and Folland's predictions of the NAO index from 1947 to 1997.[21] The upper graph shows the average predictions from an ensemble of model runs (dotted line) compared with the observed values (solid line). The shading indicates the spread of the model runs. The two lines are shown smoothed in the bottom graph. The shading here is the winter temperature in North Europe.

communicated to the atmosphere through evaporation, precipitation, and atmospheric heating processes, thereby leading to changes in temperature, precipitation, and storminess over Europe.

However, most evidence still seems to suggest that this effect of temperatures in the North Atlantic on the atmosphere is quite small compared to internal atmospheric variability on year-to-year timescales. Another possibility is that changes in tropical heating force a remote atmospheric response over the North Atlantic that, in turn, drives extratropical sea surface temperatures and sea ice. Martin Hoerling, Talyi Xuby, and Jim Hurrell at two laboratories in Boulder, Colorado, set out to see which part of the world's oceans was responsible. In order to do this, they isolated the role of tropical sea surface temperatures by replacing the observed sea temperatures and sea ice at high latitudes (outside 30°S to 30°N) in the model runs with a repeating annual cycle.[22] Their results confirmed that the NAO variations since 1950 are linked to the progressive warming and other changes of the tropical oceans, especially the Indian and Pacific Oceans. These ocean changes altered the pattern and magnitude of tropical rainfall and heating, and one of the atmospheric responses to this was the strength of the NAO. The slow, tropical ocean warming forced a commensurate trend toward one extreme phase of the NAO during the past half century.

This seems to indicate that changes in tropical heating force a remote atmospheric response over the North Atlantic that, in turn, drives changes in extratropical sea surface temperatures and sea ice. But can these results be reconciled with the North Atlantic basin calculations of the original study? If the NAO variations arise from effects of the remote tropical ocean on the atmosphere immediately above it, what is the role played by the interactions of ocean and atmosphere described by Rodwell, Rowell, and Folland? One mechanism for reconciling the two different conclusions is a climatic resonator in the Atlantic region, which is driven by climatic signals from outside.

A resonator is any object having a natural frequency, its resonance frequency. Perhaps the most familiar example is a playground swing which acts as a pendulum. Pushing a person in time with its period will make the swing go higher and higher, but attempts to push it at a faster or slower tempo result in smaller movements. Resonance occurs widely in nature and is exploited in many man-made devices, such as musical instruments. Many sounds we hear when hard objects are struck are caused by brief resonance vibrations. When the tidal wave strikes a continental shelf that is a quarter wavelength wide, a tidal resonance occurs and the incident wave can be reinforced by reflections between the coast and the shelf edge producing a large tidal range. This occurs in the Bay of Fundy, where the world's highest tides are found, in the Bristol Channel, on the continental shelves of Patagonia, and on NW Australia.

In the resonator model, the variations in the tropics constitute the external forcing and the conditions in the Atlantic the state of the resonator. The climatic resonator would then encapsulate the processes discussed by Rodwell's group. A simple model incorporating these ideas illustrates how the NAO fluctuations might arise.[23] It is made up of two components: an atmospheric component, A, representing the NAO, and an oceanic component, O, which is expected to be largely based in the Atlantic Ocean. The atmospheric component is driven by forcing from outside the region, and this component in turn drives the ocean component. At any time the accumulated condition in the ocean reacts back upon the atmosphere. Such a system will resonate for a time at some natural frequency in response to any external event.

Hoerling, Hurrell, and Xu have demonstrated that the external atmospheric forcing is largely associated with temperature changes in the tropical Pacific, Indian, and Atlantic Oceans. A readily available data set incorporating these changes is the time-series of global temperatures. These global averages will be heavily influenced by what happens in the large fraction of the earth's surface at low latitudes. There is another source of temperature variations in the equatorial region, however, and this is the ENSO phenomenon. El Niño and La Niña events have pronounced effects on tropical temperatures but, as they cause warmer and colder regions to occur at different longitudes, they may not be adequately

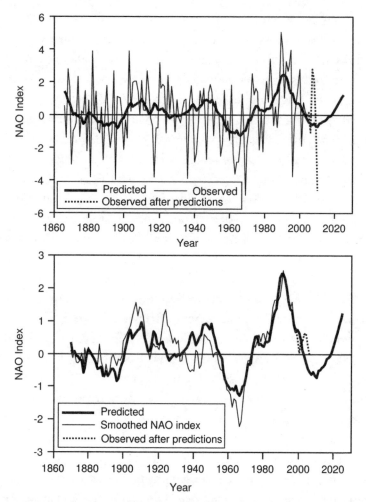

FIG 7.2 Predictions of the simple NAO model[23] (thick lines) compared with observed values (thin line). In the lower graph, the observed NAO indices have been smoothed.

represented in the global temperature data. This therefore introduces a second possible source of external forcing, ENSO cycles, which can be included in the model as the Southern Oscillation index.

Figure 7.2 shows the results obtained if each of these two data sets is used to drive a separate resonator, and the two predictions are combined.

While the NAO values in individual years show much scatter about the predicted graph, the results from this combined model do follow the general trend of the 143-year series of observed NAO values from 1866 to 2008. Just how closely they follow the trend is made clearer in the lower part of the figure in which the predictions are compared with a smoothed version of the observations. Examination of these and other calculations shows that the global temperature data provide the dominant contribution to the predictions. The climatic resonator in this case has a natural period of 44 years, while that associated with the Southern Oscillation index has a shorter period of 20 years. In the model, the oscillations were quite weak in the period up to the mid 20th century, but after 1950 they were excited more strongly, and this is the explanation for the trend in the NAO.

On this interpretation, the fluctuations in the NAO from year to year take place about an underlying trend that is constrained by the ocean.[24] It is easy to extend the model predictions into the future, provided that some forecast of global temperatures is used. Forecasts of the Southern Oscillation cannot be reliably made several years ahead but the contribution of the Southern Oscillation to the model is small enough that, for the present, it may be ignored. A forecast to 2025 is included in Fig. 7.2 based on the assumption that global temperatures continue to rise as they have done over the past two decades. It predicts relatively low values of the NAO up to about 2020, followed by a gradual rise to more positive values. This prediction depends, of course, on what is assumed to be the future rise in temperature. In the highly unlikely eventuality no further increase occurs, the model would predict a sustained period of low values extending as far as 2025. While intriguing, these are results from a very simple model. If the model does represent the dynamical balance of a much more complex system, the predictions require this representation to continue into the future. This may be possible over a couple of decades but, as global warming begins to bite over a longer period, the processes operating in nature may diverge from those in the simple model, and so any forecast becomes more uncertain.

Therefore, even though there is no clear evidence that either the North Atlantic or North Pacific Oceans strongly determine weather patterns over and beyond them, these model results and those of Rodwell, Rowell and Folland. seem to show that interactions of the ocean and atmosphere in mid latitudes still do contribute to surrounding weather in the Atlantic Ocean by generating trends in the NAO, and these changes in turn lead to effects further afield. But, even in this case, the NAO variations appear to be driven predominantly by the tropical regions, with the coupling in the North Atlantic merely providing a modulation.

The modelling work by Seagar's group has shown that the ocean accounts for about half the amelioration of western Europe's climate, the bulk of which is just the ocean storing heat and releasing it later. Northward flow of the ocean currents makes up no more than a tenth of the warming. It is likely that the sea surface temperature characteristics communicated to the atmosphere through evaporation and heat fluxes from the ocean surface in Rodwell, Rowell and Folland's model were predominantly generated by variations in heat storage, rather than by the ocean currents. The simple NAO model is similar: although the ocean component in the model falls in the 1960s as the Gulf Stream moves south before both rise in the 1990s, the changes do not agree at all closely.[23] Coupling between ocean and atmosphere in the cycling of the NAO may therefore be mainly down to heat storage and release, rather than any variations in the ocean currents.

It therefore still remains unclear why plankton populations appear to track the Gulf Stream over several decades, even though the organisms often only have short lifecycles and the ocean circulation responds to the atmosphere on a timescale of years. All the evidence presented above indicates that any signal promulgated into the weather patterns from the ocean circulation is likely to be weak. In spite of the early work by Namias, ocean temperatures do not seem to influence weather patterns strongly, and the currents of the Gulf Stream system do not play a large role in the climate of Europe, and may not do so in the NAO. In confirmation of this, attempts to isolate any European weather pattern associated with the position of the Gulf Stream have merely shown that, if there is one, it is not strikingly apparent.[25]

If this is so, is it possible that atmospheric changes caused by the ocean currents, which are buried within the natural fluctuations attributable to the global weather system, could still be exhumed by living communities? Essentially, this would require that the ecosystems effectively amplified the weak signal originating way out over the oceans. The next chapter examines this possibility by looking at how communities as a whole respond to fluctuations in the climate. There are certainly examples of the living world bringing to light things that would otherwise remain hidden, including a physical blow struck by a fist which was both invisible and intangible.

8

LIGHTHOUSES ON THE SHORES
OF THE CLIMATE

'Appearances are a view of that which is unseen.'
(Anaxagoras of Clazomenae (c. BC500–428))

In May 2007, a South Korean fisherman lifted a trap made of shells from the bottom of the sea. Inside it was an octopus the size of an orange, clutching a porcelain plate over the trap entrance in an attempt to hide. This finely glazed plate led South Korean archaeologists to a cache of up to 2000 pieces of high-quality green and blue-green porcelain from what appears to be a 12th century shipwreck. The well-preserved cups, bowls, and plates were intended for the noble class and government officials in the Goryeo Dynasty, which ruled from 918–1392AD. 'I can't believe how such a small octopus managed to cover its shell with such a large plate,' said Moon Whan-suk from the National Maritime Museum, 'I guess it meant for us to discover the artifacts.'[1] This haul came to light because of the unintentional efforts of one animal but sometimes a population is involved.

In the 1970s, a geologist discovered a single fleck of a mineral called ilmenite on the surface of the Kalahari desert. Ilmenite comes from a type of rock called kimberlite, which hosts diamonds, and the fleck revealed the richest diamond deposit in the world at what is now the Jwaneng diamond mine.[2] The minerals were 40 m down, and the grain would not have reached the surface but for termites. Desert termites dig deep. In hot, arid areas they build large mounds above ground to help air circulation and temperature control, and, if these need repair, the insects tunnel 30 m or more down to get the wet mud they require. Even though termites can give a nasty bite, and their mounds may house venomous snakes, mineral prospectors often look into their diggings for telltale signs of deposits such as gold. These minerals are brought into the sunlight by the activities of a single insect population, but the impact of one much larger blow was only revealed by changes in a whole community.

Joseph Nkwain was awakened at about midnight by a loud noise:

'I could not speak. I became unconscious. I could not open my mouth because then I smelled something terrible…I heard my daughter snoring in a terrible way, very abnormal…When crossing to my daughter's bed…I collapsed and fell. I was there till nine o'clock in the (Friday) morning…until a friend of mine came and knocked at my door…I was surprised to see that my trousers were red, had some stains like honey. I saw some…starchy mess on my body. My arms had some wounds…I didn't really know how I got these wounds…I opened the door…I wanted to speak, my breath would not come out…My daughter was already dead…I went into my daughter's bed, thinking that she was still sleeping. I slept till it was 4:30 p.m. in the afternoon…on Friday. (Then) I managed to go over to my neighbors' houses. They were all dead…I decided to leave…(because) most of my family was in Wum…I got my motorcycle…A friend whose father had died left with me (for) Wum…As I rode…through Nyos I didn't see any sign of any living thing…(When I got to Wum), I was unable to walk, even to talk…my body was completely weak.'[2]

Joseph Nkwain was a survivor of a sudden disaster near Lake Nyos in the western Cameroon, adjacent to Nigeria. At 9.30 p.m. on August 12, 1986, a lethal mist swept down, like a biblical plague, killing over 1700 people, thousands of cattle, and many more birds and animals. There was no accompanying physical disturbance: no houses were damaged, and no trees or branches were knocked down. The bodies of those that died were generally devoid of trauma. Most victims appeared to have simply fallen asleep and died from asphyxiation, and many died in their beds. Local villagers attributed the catastrophe to the wrath of a spirit woman of local folklore who inhabits the lakes and rivers. Scientists, on the other hand, were initially puzzled by the root cause, and by the abrupt onset, of this mysterious and tragic event.

They eventually discovered that the killing agent was a cloudy mixture of carbon dioxide and water droplets that had risen violently from the lake. The red stains and sticky coating may have been washed off trees by the droplets. It had been known for years that the water in Lake Nyos was extremely enriched in dissolved carbon dioxide. The lake overlies a volcanic source, which appears to release this and other gases. Most of the gas does not escape into the atmosphere, but rather dissolves into the bottom waters of the lake. At a depth of over 200 m, the sheer weight of the upper lake levels exerts considerable pressure on the bottom waters (about 20 atmospheres) and this confining pressure allows the carbon dioxide to dissolve into the bottom waters without escaping to the surface, in much the same way that the cap on a fizzy drink prevents carbon dioxide from bubbling out of its container. Under these conditions, water can hold up to five times its own volume in carbon dioxide, so that every litre of water in the lower part of the lake could have contained 1–5 litres of carbon dioxide.

On that fateful August evening, this gas was released when the lower layers of the lake were somehow brought up to the surface and it is thought that the rapid accumulation of rainwater in the lake was responsible for mixing the bottom waters to the surface. The rainwater may have been blown to one side of the lake by strong August winds and, being denser than the warmer lake water, would have descended down one side of the lake, displacing the bottom waters. The resulting overturn led to the ascent and decompression of the water from the deep, whereupon the dissolved gas came out of solution and bubbled upward at dramatic speeds. The rapid decompression also dramatically cools the gas so that it causes frostbite in those exposed to it. As this happened, the bubbles themselves may have lowered the overall density of the gas–water mixture, causing even greater rates of ascent, decompression, and expulsion. A rapid and violent outgassing of carbon dioxide ensued, and so much gas escaped from this single event, that the surface level of the lake dropped by a metre. To prevent such an event happening again, large pipes have now been placed into the lake which siphon water continuously from the lower layers to the surface and allow the carbon dioxide dissolved in the bottom waters to slowly bubble out.

Lake Nyos' gigantic belch was a major tragedy to the local people and their livestock but, leaving no physical traces in the landscape, it was also a dramatic event that would not have been observed but for the presence of people and other animals. In their absence, it might have passed unnoticed. While this is an extreme example, there are other cases in which biological populations or ecosystems highlight changes or features in the environment. Recently, the remains of the world's largest known snake were unearthed from rocks that were 65 million years old in a coal mine in north-eastern Columbia by a team of palaeontologists lead by Jason Head at the University of Toronto. The animal, at 13 metres, was longer than a city bus, had a diameter of a metre and, with a weight of more than a tonne, was heavier than a bison. From comparisons with living snakes, the group estimate that mean annual temperatures of 30–34°C would be required to support such a snake. This is because these animals have internal temperatures varying with the ambient temperature and need to sustain a minimum metabolism to survive. Temperatures this high are above those tolerated by modern forests, and so, if the snake was like its modern counterparts, the forest must have been unlike those we know today. This fossil animal therefore provides us with 'a window to a time and place that we've never had access to before.'[3]

Fossil animals and plants have long been used in this way. In Chapter 3, William Smith was described walking along a Roman road on his way to construct the first geological map, drawing on his knowledge as a surveyor for the Somerset Coal Canal. When he came to do this, he needed some way of telling that the carboniferous rocks from Devon are younger than the Cambrian rocks from Wales. The solution came to him on the evening of the 5 January 1796 while sheltering from the cold in a coaching inn in Somerset: the answer lay in fossils. As rock types changed with time, some species of fossils disappeared while others continued on into subsequent levels. By noting which species of fossils occurred in which strata, Smith could work out the relative ages of rocks wherever they appeared. The arrangement of fossils enabled him to predict what was underground where, and thus to construct a map of Britain's rock strata. In 1815 it was the first of its kind.[4] Smith's map and its successors depend on the demise of living creatures to reveal and mark out past geological events.

But organisms do not have to be fossilised to highlight environmental processes. Aerial photographs show up archaeological sites as contrasts between light and shadow, and differences in the appearance of the soil, but especially by differences in height and colour of cultivated crops. Colour changes and faint lines visible from the air, but invisible on the ground, can be caused by buried cultural remains. These different visual indicators are referred to by aerial archaeologists as shadow-, soil-, and crop-marks, respectively. Crop-marks are formed when buried archaeological features cause differential crop growth. As crops begin to ripen in early summer, the walls, pits, and ditches of past settlements affect the rate at which crops change colour, and also the speed and height to which they grow. For example, crops growing over a buried ditch will be taller and larger than those in surrounding soils, since the ditch often contains additional moisture. Crops growing over the buried remains of a wall, which encourages water to drain away from the soil and interrupts root growth, will tend to be smaller. The appearance of crop-marks is enhanced by dry conditions, such as during times of drought, when differences in the ripening crop's condition can become very noticeable in aerial photography. Certain crops such as wheat, barley, and other cereals produce crop-marks with especially good definition and resolution.

Ancient settlements can be discovered even more efficiently if chemical analysis is used as well. Plants growing over old sites of human habitation have a different chemistry from their neighbours and these differences can reveal the location of buried ruins. Thus, when Rob Commisaro and

Eric Nelson from Simon Fraser University in British Columbia, Canada spent three summers collecting plants from sites in south-west Greenland, they found some of their samples were unusually rich in the isotope nitrogen-15, compared to the more usual nitrogen-14. Subsequent digs revealed that these plants had been growing above long-abandoned Norse farmsteads.[5] The enrichment occurred because over archaeological sites the nitrogen was derived from refuse or other nitrogenous compounds that people had deposited in the past.

This kind of chemical analysis can even be carried out remotely. Hidden graves can be detected by using airborne imaging to detect changes in vegetation too subtle to be seen by the naked eye. Cameras mounted on a light aircraft or helicopters observe variations in the intensity of light at various wavelengths reflected by vegetation on the ground. The precise pattern of intensities has been found to reflect changes caused by nutrients released into the soil as bodies decompose.[6]

Perhaps the most well-known example of living populations highlighting events is the widespread belief that animals can predict earthquakes.[7] In 373 BC, historians recorded that animals, including rats, snakes, and weasels, deserted the Greek city of Helice in droves just days before it was devastated by an earthquake. There are more recent examples. In 1975 the Chinese authorities ordered the evacuation of Haicheng, a city with one million people, shortly before a magnitude 7 earthquake struck. Animal behaviour gave the warning and then a series of tremors called foreshocks provided the Chinese officials with a more solid prediction. This incident prompted the US Geological Survey to look at animal behaviour as an aid to earthquake prediction. What the animals may sense is a mystery, possibly they feel the Earth vibrate before humans, or maybe they detect electrical changes in the air or gas released from the ground. However, animals react to so many things: hunger, defending their territory, mating, or predators that the USGS found it hard to do a controlled study, and nothing came of the investigation. Although the possibility of animals forecasting earthquakes is still pursued in vulnerable areas like China and Japan, the idea remains controversial.

A recent observation suggests that toads at least may sometimes be aware of an impending earthquake. Rachel Grant of the UK's Open University monitored a population of reproductively active common toads *Bufo bufo* over a period of 29 days, before, during, and after the earthquake (on day 10) at L'Aquila, Italy, in April, 2009. Although her study site is 74 km from L'Aquila, toads showed a dramatic change in behaviour 5 days before the earthquake, abandoning spawning and not resuming normal behaviour until some days after the event. This is most unusual. Breeding sites are male-dominated and the toads would normally remain in situ from the point that breeding activity begins, to the completion of spawning. It is unclear what environmental stimuli the toads were responding to so far in advance of the earthquake, but reduced toad activity coincided with pre-seismic perturbations in the ionosphere, detected by very low frequency radio sounding.[8]

The effects of climatic variations on biological populations are likely to be especially strong in the far north where conditions are harsh and variable. Satellite observations have shown sea ice coverage in the Arctic to be shrinking at about 8% per decade since 1979, and in 2001 its thickness had reduced to an average of 1.8 m compared to 3.1 m in the 1950s. Field observations from the Beaufort Sea to Hudson Bay suggest that many species are floundering in this warming environment. For example, Inuit hunters living on the northernmost fringes of Canada have reported that Ivory gulls, birds that live in remote areas far from prying eyes, are disappearing. Seabirds tied to cliff-side nest colonies can forage only as far northward as there is suitable land and so, for some bird species, the balance between ice and open water can be critical for survival. Black guillemots off Alaska's northern coasts are a species that illustrate the subtleties of this balance.[9] Before the 1970s summer cold and snow prevented these birds from nesting on Cooper Island, in the Beaufort Sea near Point Barrow, Alaska, but George Divorky of the University of Alaska found that, as the seasonal ice edge pushed north, the birds settled and flourished with numbers peaking in 1989. However, in the next decade the population halved. As the ice receded even further north, the ice edge outran the guillemots and is now far out in the Chuckchi-Beaufort Sea where there is no land. The birds have to commute too far to reach the arctic cod on which they feed, and their chicks either starve or are eaten before they return. As a result, the black guillemot is now locally extinct on Cooper Island.

These conditions are specific to the seabirds of the high arctic, but there are some general features of populations that can make communities particularly susceptible to the vagaries of the climate. One such is the mistiming that can arise when the development of one population is synchronised with a particular event. An example of this is the *Daphnia* population of Windermere, described in Chapter 3, where the animal has its growth physiologically timed to match the appearance of the phytoplankton on which it feeds, and so suffers if the algae have been and gone earlier. The to-ing and fro-ing of migrating birds may often make them vulnerable to climatic events happening at an unusual time. In a study of pied flycatchers,[10] Christiaan Both and colleagues at the Netherlands Institute of Ecology found that, over 17 years, pied flycatchers declined strongly in areas where caterpillar numbers (food for nestlings) peaked early, the young birds arriving too late in areas where caterpillars had responded to early warmth. Meanwhile, in areas with a late food supply, there was no decline. Mistiming like this may well be a common consequence of climate change, and may be a major factor in the decline of many other long-distance migratory birds.

Of course, the mistiming simultaneously benefits the algae or the caterpillars that are being preyed upon, and such indirect interactions may be crucial to how a community responds to environmental changes. For instance, because ultraviolet light inflicts damage on algae in streams, removing UV might be expected to enhance primary production. But this is not the case: it often turns out that the insects grazing on the algae are even more susceptible to burning by UV, and so sheltering streambeds from these rays enhances the insects' survival, reducing algal populations.[11] Environmental stress on one set of organisms can therefore propagate through the food chain and, in Chapter 4, it was shown that indirect interactions may even be so minutely sensitive to environmental conditions that communities starting from very similar initial conditions may follow disparate paths, ending up with wildly different outcomes. One particular situation in which indirect interactions make a community particularly sensitive to surrounding events is when one species or group of species plays a key role in the ecosystem.

An example of this situation has been provided by experiments on the tiny ecosystems that flourish in carnivorous pitcher plants.[12] Insects such as flies are trapped by the water in the plant's bowl, and then become fertilisers for the plant. The conversion from animal body to nutrients depends upon the tiny community present in the water. Nick Gotelli and A.M. Ellison at the University of Vermont have used these tiny systems to show how the most important interactions in a community can be extracted and used to predict its dynamics. In their experiments, they manipulated the volumes of their pitcher-plant habitats and varied the abundance of dipteran larvae (the most important predators in the pitcher plants). They then measured the responses of the aquatic mites, rotifers, protozoans and bacteria living there to these changes. Gotelli and Ellison found that the density of the larvae was critical to the food web. The effects on the community of shrinking the size of the habitat could be mainly explained as the indirect consequence of its impact on this top predator, which requires a substantial volume to support it, and so is most likely to be affected by any loss of habitat.

These experiments offer an important lesson: even though small in numbers, the top predators often exert a disproportionately strong influence over the communities in which they live. This observation is not unique to the ecosystems in pitcher plants; bromeliads in Costa Rica tell a similar story. Water held inside the leaves of these plants provides an environment similar to those of the pitcher plants. Here, any increase in the number of leaves reduces the foraging efficiency of predatory damselfly larvae, and this change has a substantial effect on the water's communities.[13] In each of these two systems reduced predation caused by changes in habitat percolate down through just a few strong community links, such as the interaction between bacteria and their single-celled predators.

But it is not just the tiny aquatic ecosystems in plants for which the top predator's activities are crucial; the loss of such predators as jaguars from tropical forests on Venezuelan islands following habitat fragmentation has also had community-wide results.[14] The islands were formed when a valley was flooded to develop a hydroelectric scheme in the 1980s and a lake, Guri, was created. Many plants that could easily cope with the loss of habitat still felt the impact of the flooding because the number of herbivores was not held in check by carnivores. Smaller islands, on which there were almost no predators of the vertebrate grazers, and the armadillos that prey on the leaf-cutter ants were absent, subsequently suffered a pronounced decline in vegetation. William Stolzenburg has argued that the presence of top predators is vital for many ecosystems.[15]

In all communities like these in which the presence of the top predator has such a profound effect, any weather changes impacting on the predator will have widespread effects on the ecosystem. A species that has a disproportionate effect on its environment relative to its abundance is referred to as a 'keystone species'.[16] It is analogous to the keystone in an arch. While the keystone feels the least pressure of any of the stones in an arch, the arch will still collapse without it. Similarly the ecosystem may experience a dramatic shift if the few members of the keystone species are removed or their number is changed. Among the many examples of such species is the grizzly bear on the west coast of North America. These bears feed on Pacific salmon that swim hundreds of miles up the rivers to breed. The bears carry the salmon onto dry land, dispersing nutrient-rich faeces and partially eaten remains into the forests. It has been estimated that the bears leave up to half the salmon they harvest on the forest floor. If the bear populations falter, then so will many other parts of the forest community.

Another keystone species is the Californian sea otter, which was hunted almost to extinction for its valuable pelt.[17] In places where the sea otters disappeared completely, sea urchins, normally among the major prey of the otters, exploded in numbers and proceeded to consume the kelp which is anchored to the seabed and reaches to the surface as a forest. Few fish could live in the barren areas produced and so great fisheries and canneries along the coast disappeared just as the sea otters had done. In 1960, Nelson Hairston, Fred Smith and Larry Slobodkin in the USA even proposed that the world is green because predators keep herbivores under control and allow plants to flourish. However, as Lauri Oksanen and his colleagues pointed out 20 years later, only in the most productive ecosystems would greenness prevail because of predators holding herbivores in check.[18] Where plant production is low, in tundras or deserts, the few transient herbivores eat so little overall that their removal would not be noticed. In more productive systems like grasslands, more abundant plants can support effective grazers that can strip the greenery but are not eaten by larger predators in significant numbers.[16]

It is not always the top predator that dominates a community. The Canary Islands, situated only 70 kilometres west of the Sahara desert, receive virtually no rainfall. Even so, when Spanish sailors landed on one of the islands in the 15th century, they found a local people, the Guanche, with an extensive agriculture. A major explanation for this seems to have been the trees: moisture-rich fogs drift in from the Atlantic Ocean, and the water dripping from the leaves was sufficient to support the people. At the time of the Guanche, all seven islands had rich forests. Since then, much of the islands' forests have been cut down for firewood, construction and to make way for farmland. While most of the islands still have some degree of forest cover, one, Lanzarote, is almost bare. With the loss of forest, the drying out of the land destroyed the ecosystems and all efforts at reforestation have so far failed. Water supplies on the island are insufficient to prevent newly planted trees from drying out. Although clouds of moisture form over the mountains of northern Lanzarote even during the hottest months, the surface of the mountains are still too hot for this to happen at ground level, and so the fog hovers out of the reach of the saplings. The tall trees were essential to the lost ecosystem.[19] Even so, technically the trees still do not qualify as a keystone species. This is because their biomass is large and their impact is not disproportionate to it.

Conditions on Lanzarote were changed dramatically when the tranquil island was shaken by a series of volcanic eruptions that continued almost without a break from 1730 to 1735.[20] The eruptions sprayed the landscape with black volcanic stones that triggered an agricultural revolution. Before them the farming had been rudimentary and the population poor, but afterwards the land was transformed from near desert into a fertile oasis. The farmers had noticed that, in the areas that had become blackened by the stone, crops didn't die but frequently grew between the stones with renewed vigour. Apart from shading the soil from the sun and so reducing evaporation, the pumice-like stones are porous and trap moisture, as well as stopping weeds growing and making ploughing unnecessary. Formerly useless hillsides were cleared of brush, cut into terraces, covered in the stones—which the local people called picon—and planted with vegetables and other crops. Grapes did particularly well if the vines were sheltered against the year-round trade winds by creating shallow pits and forming wind-breaks out of the stones. Later the farmers learned to grow prickly pears, not for the fruit of the cacti but for the cochineal insects that infested them. The toxic carminic acid that the insect uses to deter predators provides a crimson dye, cochineal, today used in everything from lipstick to strawberry milkshake.[18]

Elsewhere in the middle of the North Atlantic the forest changes that took place in the Canary Islands have happened in reverse.[21] When Charles Darwin called in at Ascension Island on board the *Beagle* in 1836, he found the island to be entirely devoid of trees with a bare white mountain in the middle. All it had was a few species of plants, mostly ferns, some of which were found nowhere else. The reason was that the island was young—about a million years old—and, at 2000 km from the nearest continent, was not easily colonised. During the middle of the 19th century, the Royal Navy, which maintained a garrison on the island, planted an assortment of plants, including two that did particularly well: bamboo and prickly pear. The mountain was soon renamed Green

Mountain. An Admiralty report in 1865 described how the island 'now possessed thickets of up to 40 kinds of trees besides numerous shrubs,' and noted that 'through the spreading of vegetation the water supply is now excellent'.

A species or a group of species in an ecosystem may still be especially sensitive to environmental change even without being a keystone species or being vital to all the rest of the community. Ecologists Alan Knapp and Melinda Smith from Kansas State University came across an example of this when they used data from the US National Science Foundation's Long Term Ecological Research (LTER) sites to investigate how three biomes: arctic tundra, grasslands and eastern forests were affected by precipitation.[22] At the 24 North American LTER sites, researchers continue to monitor precipitation and to estimate aboveground net primary production by carefully measuring the growth of plants each year. Because the measurements are taken in a consistent way at the various sites, they can be used to compare the different ecosystems. As might be expected, productivity was higher at sites with more annual rainfall. Forests, with their relatively huge plants, had the highest production, with grasslands coming second and deserts third.

However, Knapp and Smith found a different pattern in how each of these biomes responded to fluctuations in precipitation. Forests, which tend to receive fairly stable amounts of rainfall, grow roughly the same in wet or dry years. In addition, deserts, some of which experience the wildest swings in rainfall, still fluctuate only moderately. The biomes that were most sensitive to rainfall proved to be grasslands, these being four times more so than forests. Grasslands may be so variable because of their underlying growth potential: they have more leaf area and can grow more densely than plants in deserts. Forests receive and retain much more water and so are more buffered against dry years. Knapp and Smith also found that in all the biomes wet years had a much greater effect on plant growth than did dry spells. They attributed this to plant features that enable them both to resist droughts and to sprout new growth when well watered, a resilience that had been noted by plant physiologists.

When reporting Knapp and Smith's results, the journal *Science* described grasslands as an ecological bellwether. (The term bellwether, an entity that serves to influence future trends or presage future happenings, is derived from the ancient practice of placing a bell around the neck of a castrated ram, a wether, in order that this animal might lead its flock of sheep). If so, the growth potential of grasslands may mean that they are indicators and may be early warning signs of environmental damage and ecosystem change. This importance of grass is not an entirely new idea: its crucial significance had been described 50 years earlier in a fictional context. John Christopher, in his novel *Death of Grass*, examined what might happen if a virus were to wipe out all grasses. In this story, the resultant collapse of ecosystems leads to such famines that governments resort to having the military kill off some of their own population.

Ecologists have identified bellwether species in various ecosystems. One is the Attwater prairie chicken, which once numbered in the millions in Louisiana and along the Texas coast but is now almost extinct. Another is the northern diamondback terrapin, which conservationists are trying to save along the Massachusetts' Outer Cape. Commercial fishing is endangering northern and southern petrels, large scavenger birds of the Antarctic Peninsula, and human impacts are causing trouble for polar bears in the Arctic. Each of these species is considered to be an extreme indicator of problems within its ecosystem.[23]

Bellwether species might be sensitive indicators of climatic fluctuations even without being on the verge of extinction. Windermere's water fleas (see Chapter 3) may be just such a case, the *Daphnia* being especially sensitive to the prevailing weather because of their need to have their development in time with their algal food supply. The roadside verge at Bibury, whose monitoring was discussed in the same chapter, may be another example. Perhaps, the field scabious shown in Fig. 3.2 might be considered a bellwether species for distant changes in the ocean. This is a grassland site in which competition between different types of plants is important, and so it is unlikely that a single ecological bellwether will explain how a weak weather effect from the circulation of the North

Atlantic Ocean appears in the community. Neither can it account for the Gulf Stream relationship in the plankton of the North Sea. Here the total number of copepods is made up of several kinds of animals with differing lifestyles. There is another way that ecosystems could be sensitive to climatic variations, however: separate parts of the community might respond to different aspects of the weather changes, and the community as a whole can then accumulate these impacts. There are examples of this process in action. One of these is the cycle of ice ages.

Understanding how ice ages wax and wane came from the efforts of perhaps the most contented prisoner of war in history, Milutin Milankovitch.[24] But he wasn't the first to tackle the question. When it was discovered in the 19th century that the earth had been subject to a succession of ice ages, there was then a need to find out what caused them. A possible explanation was provided by a man employed at Andersonian College and Museum, Glasgow, working, not as one of the academic staff, but as a janitor. James Croll had come from a poor background with a formal education that finished at age 13, and had taught himself by reading in the university library during the evenings. His situation changed when he published a paper proposing that the repeating ice ages of the earth are attributable to slow variations in the earth's motion around the sun. Following this and other papers that he wrote, Croll was given a job with the Geological Survey of Scotland, eventually receiving many honours.

Croll's idea was that decreases in winter sunlight would favour snow accumulation and this effect would be amplified by the extra ice reflecting more sunlight into space. Based on formulae developed by Leverrier when he predicted the existence of the planet Neptune, Croll realised that the Earth's orbit changes shape in a periodic fashion. He suggested that, at times, the eccentricity of the orbit is high, so that the orbit is less circular and more elliptical, then winters will tend to be cold when the Earth is farther from the Sun. However, his theory predicted multiple ice ages would occur but they would be out of phase in the northern and southern hemisphere. When the orbit had the northern winter happening further from the sun, the southern winter would occur closer. The theory also predicted that the last ice age should have ended 80,000 years ago, when in reality it was much more recent.

Croll's theory was picked up and extended in the early 1900s by Milutin Milankovitch, a Serbian mechanical engineer. Milankovitch realised that the gravitational attraction between the Earth, the Sun, the planets, and especially the Moon, generated rhythmical shifts in the shape of the Earth's orbit, also altering the planet's angle of orientation to the Sun, affecting its tilt, pitch and wobble. These all change the intensity of sunlight falling on the surface in any season and lead to cycles of 23,000 years, 41,000 years, and 100,000 years. He set about working out how much solar radiation reached each latitude on the Earth in every season over the past million years. Computing the tables of these cycles was to involve him in 20 years of ceaseless effort. When World War I broke out he was arrested because he was a reservist in the Serbian army. He spent most of the war under loose house arrest in Budapest, required only to report to the police once a week, and so his status as a prisoner of war left him ideally placed to pursue his calculations in the library of the Hungarian Academy of Sciences.

Milankovitch died in 1958, but it was not until the 1970s that refinements in dating ancient seafloor sediments allowed his theory to be tested. By then it was understood that the key to explaining how orbital variations can trigger ice ages is the amount of solar radiation received at high latitudes during the summer. This is critical to the growth and decay of ice sheets. The ice sheets spread further when winter snow cover does not melt properly during the summer months. It is now accepted that the latitudinal and seasonal fluctuations in incident solar radiation due to precession of the equinoxes (the 23,000-year cycle) and the variations in the tilt of the Earth's axis (the 41,000-year cycle) are sufficient to drive the climatic changes observed on these timescales. The 100,000-year cycle in the eccentricity of the Earth's orbit considered by Croll is the weakest of the orbital effects, and yet this cycle is the strongest feature in the observed climatic record over the last 800,000 years. How does this happen?

It seems that this comparatively weak forcing is amplified in some way by the processes within the geosphere and biosphere. The Earth's climate system may have a natural frequency with a period of 100,000 years which is then excited, much as a bell naturally rings at a certain pitch. Several features of the ice-sheet dynamics have been proposed for this, but none has been agreed upon. In the 1980s it was suggested that random noise in the form of short-term climatic fluctuations, if strong enough, could cause Earth's climate system to flip between two modes of operation: cold glacial periods and warm interglacial periods, a process called 'stochastic resonance'.[25] However, in recent years, scientists have turned instead to changes in atmospheric carbon dioxide as a possible explanation. Carbon dioxide concentrations from the past can be measured in ancient air bubbles preserved in sequences of cores drilled into the Antarctic ice sheet, and some of the observed changes have been found to occur slightly before changes in ice volume. A careful analysis of oxygen isotopes by Nick Shackleton at the University of Cambridge led him to conclude that carbon dioxide is an additional independent 'driver' of the size of ice sheets, along with the oscillating solar radiation.[26]

The deep-sea sediment oxygen isotopic composition, which is a measure of the volume of the ice sheets, is dominated by the 100,000-year cycle and Shackleton used an ice core from Vostok in the Antarctic to separate the oxygen isotope signals of ice volume and the deep-water temperature throughout it. He found atmospheric carbon dioxide, air temperature, and deep-water temperature were each in phase with orbital eccentricity, whereas ice volume lagged behind these three variables. He concluded that the 100,000-year cycle did not arise from dynamics of the ice sheet; instead, it was probably the response of the global carbon cycle generating changes in the atmospheric carbon dioxide concentration as a result of the varying sunlight.

Although there is as yet no broadly accepted explanation for the differences occurring in ice ages, many current investigations focus on the oceans' 'biological pump', the burial of carbon that rains downwards out of the sea surface.[27] The biological pump is the process by which CO_2, fixed in photosynthesis is transferred to the deep ocean as dead phytoplankton cells, or the remains of those that get eaten sink from the surface. Perhaps the reservoir of algal nutrients was larger during glacial times, strengthening the biological pump at low latitudes where nutrients are scarce. Alternatively, the biological pump may have been more efficient during glacial times because of more complete utilisation of nutrients at high latitudes where much of the nutrient supply currently goes unused. Whatever the mechanism actually is, it seems that the planet's ecosystems may well have been amplifying the weak 100,000-year signal. It is somewhat reminiscent of the feedbacks in the Gaia hypothesis, but unlike the Gaia scenario in which the global ecosystem of the Earth continually operates to maintain an unchanging environment, in the present case it has been operating to accentuate weak shifts in the environment.

Evidence of ecosystems responding as a whole in a sensitive way to weather changes is not restricted to geological timescales, nor to wild communities; the phenomenon has been seen in agricultural populations. In 1994, Mark Cane from Lamont Doherty Observatory of Columbia University and colleagues showed that forecasts of ENSO cycles could be used to predict maize yields thousands of miles away across the world in Zimbabwe. Maize is Africa's favourite crop, being the most important food crop for the ten-nation Southern African Development Community region. The crops are very dependent on rainfall, and southern Africa is subject to recurrent droughts which cause severe food shortage. There is considerable evidence that some of these may be linked to El Niño events in the Pacific Ocean. The 1991–92 El Niño event, in particular, was accompanied by the worst drought of the century, affecting nearly 100 million people. In a paper in the journal *Nature*,[28] Cane's group showed that there was a strong correlation between a measure of the status of the ENSO and both rainfall and maize yield in Zimbabwe (Fig. 8.1). They found that more than 60% of the variations in the yield could be accounted for by sea surface temperatures half-way around the world in the eastern equatorial Pacific Ocean, and that model predictions of El Niño provide

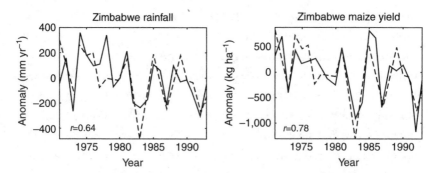

FIG 8.1 Rainfall and maize yields in Zimbabwe (solid lines) compared with predictions based on an index of the El Niño cycle. The correlation coefficient, *r*, measures how good the prediction is, showing that it is better for maize than rainfall.[28]

accurate forecasts of maize yield in Zimbabwe, with lead times of up to a year. African farmers who use weather forecasts based on ENSO data are likely to see an increase in crop yields because the accurate rainfall forecasts allow them to plant high-yielding varieties of maize when appropriate. A strong El Niño can cut maize yields by 50% in southern Africa and, if the global climate were to change to become more El-Niño-like, African food production could be severely reduced.[29] There is a glimmer of hope, however, because some crops such as sorghum can do better when maize does badly.[30] Predictions of El Niño events can therefore provide valuable information for the farmers.

A surprising result from the forecasts is that the direct correlation of the ENSO with maize yield is stronger than that with the total precipitation, i.e. the predictions of the plant yields were considerably better than those of the rainfall on which the growth depends. It was expected that the correlation with rainfall would be the stronger because growth rates depend on other factors, such as pests or diseases, which may be unconnected with the ENSO. In some way the maize was amplifying the meteorological signal. Crop productivity (both the quantity and quality of the yield) depends on a complex combination of climate, biophysical factors, and management. It seems the connection with El Niño is not adequately described by a single climate variable, total precipitation, but instead involves a wider range of variables.[31] If so, and weather patterns originating from processes in the distant Pacific Ocean can be more clearly manifested in agricultural crops than in direct meteorological observations, then perhaps atmospheric effects deriving from the North Atlantic have been similarly appearing in the ecosystems around the North Sea.

Icarus Allen and Paul Clark—from the UK's Plymouth Marine Laboratory—and I decided to investigate this possibility with a state-of-the-art model of the ecosystem in the North Sea. The model concerned, the European Regional Seas Ecosystem Model (ERSEM), has been developed in a collaborative venture between scientists from many European countries, and it attempts to take account of as many of the interactions within the plankton community as possible. It also incorporates a model of the formation and breakdown of thermal stratification in the water column. When we subjected the combined model to three decades of daily weather observations made on the Irish mainland, a location that was between the Gulf Stream and the North Sea, the plankton populations it generated rose and fell according to the position of the Gulf Stream in much the same way as those from the North Sea were observed to do.[32] The main signal to be seen in the weather observations we employed was the ups and downs of the North Atlantic Oscillation, but these were less clearly registered by the plankton, perhaps because the NAO is mainly a feature of the winter and is much less represented in spring and summer when the plankton are growing.

Examination of the results in more detail showed that individual members of the plankton were responding to different aspects of the weather. For example, winter phytoplankton populations picked up the Gulf Stream signal from how the amount of cloud cover in this season went up and down from one year to the next, while the phytoplankton in spring got it through the vernal wind strength fluctuations. When all the members of the ecosystem interacted, differences in the particular aspects of the weather tended to cancel out and a common response, traceable to the distant ocean, emerged. As mentioned in the opening chapter, something similar seems to have happened in the zooplankton of the North Sea. The linkage with the Gulf Stream is seen much more clearly when the different species of copepods are totalled together than if the species are examined individually. Each of these cases is analogous to the way that aerial photography aids archaeology by looking at the vegetation as a whole and merging the details of all the individual plants on the ground.

An interesting extra result obtained from the model is that the tendency of the plankton to track movements of the Gulf Stream might not be a permanent feature of the community. It seemed to happen for more than 20 years after which the effect suddenly switched off. There is evidence of this also occurring in some of the plankton observations. Perhaps this result is a warning that such climatic phenomena associated with the ocean may also be transitory.

Not only may the responses of ecosystems sometimes be transitory, but also there is the opposite possibility that we could be increasing the sensitivity of populations to climatic variability. In the late 1970s, theorists suggested that, as fishing depletes fish stocks by removing fish from the sea, it also reduces the resilience of fish populations in the face of change. Discussion of this topic remained theoretical until a recent analysis of the 50-year larval fish survey of waters off California. The survey, as mentioned in Chapter 1, was begun when sardine populations collapsed in the 1940s and has been maintained by the California Cooperative Oceanic Fisheries Investigations (CalCOFI). When Chih Hsieh of Scripps Institution of Oceanography in California and colleagues studied these data, they found that exploited species exhibit higher variability in abundance from year to year than unexploited species.[33] They concluded the increased variability of exploited populations is probably caused by the tendency to remove the oldest and largest fish. Many fish use strategies to hedge their bets and increase the survival rate of larvae under harsh environmental conditions, and such strategies are often associated with the older individuals in the populations. Older, larger, and perhaps more experienced, fish may spawn in different locations and at different times, and may produce more and better quality eggs. But, as has been described for the pike of Windermere in Chapter 3, these are just the animals most likely to be fished out. Selective removal of the largest and oldest fish amplifies the peaks and valleys of population size because it reduces the capacity of populations to buffer environmental effects.

This is not the only example of such indirect effects of human activities. Plants and soil are currently slowing down global warming by storing about a quarter of human carbon dioxide emissions, and this mitigating effect could be undermined by increases in ozone. Unlike high altitude ozone, which blocks harmful ultraviolet rays, near-surface ozone damages plants, reducing crop yields and the capacity of plants to take up carbon dioxide. Ozone is formed at ground level in a reaction between sunlight and gases such as nitrogen oxides, methane, and carbon monoxide. Stephen Sitch and colleagues from the UK's Hadley Centre have explored this effect on plants by using a global land carbon model which was modified to include the effect of ozone on photosynthesis and to account for changes in the closing of the stomata of plants. They found that projected increases of ozone concentration led to a significant suppression of the global carbon storage on land, and further that this indirect effect could contribute more to global warming than the direct effect of ozone as a greenhouse gas.[34] Rising ozone pollution over coming decades could therefore make the planet more susceptible to carbon emissions.

Even without this kind of human help, the earlier examples illustrate several ways ecosystems might be susceptible to quite subtle changes in weather patterns. The weak signal may be amplified because organisms show abrupt changes in response to relatively small changes in weather patterns, as happened to the black guillemot. This is what happens when there is a mistiming between an animal and its food, such as occurs with the *Daphnia* in Windermere and, as noted earlier, is something to which migrating species are prone. A community may also be strongly affected by a series of weather events if they impact on a key member of the community to which all are connected. This may be a dominant species, as in the trees of the Canary Islands, or merely a keystone species, as is the case with the top predator in some ecosystems. Alternatively, an individual species may just be a bellwether species, highlighting changes in climate but having only a weak impact on the rest of the ecosystem. Finally, a community may respond in a more holistic way to different aspects of the changing weather patterns: several members of an ecosystem may be affected by an underlying climatic signal which then becomes a common response as other more individual weather effects cancel out in the interactions of the system. In the case of the maize in Zimbabwe, the effects of El Niños may have been brought out from weather events originating elsewhere by a combination of ecological interactions and crop management strategies. The modelling experiments with ERSEM suggest ecological interactions are what causes zooplankton in the North Sea to be linked to the position of the Gulf Stream.

If ecosystems do sometimes amplify weak changes in weather patterns in these ways, they have the potential to provide an early warning of the impact of climatic changes. Monitoring the behaviour of the community will then serve a very similar purpose to maintaining a lighthouse near the sea. As the light gives a warning of hidden dangers below the sea's surface to approaching ships, so changes in the populations and their interactions may warn of early shifts in weather systems that might otherwise pass undetected and might be a portent of what lies ahead.

The need to observe communities that are undisturbed means that monitoring programmes are often sited in almost inaccessible locations of great beauty, and this is the case with many lighthouses. One such lighthouse had an unfortunate impact on its local wildlife. In 1894, Tibbles, the cat of the lighthouse keeper on Stephens Island, New Zealand started catching a small, flightless bird. By the time the keeper had found out that this was the sole-remaining population of an unknown species, now called the Stephen's Island Wren and which had previously been widespread across the whole of New Zealand, Tibbles had made the species extinct.

The analogy between monitoring programmes and lighthouses goes even further: both have been frequently set up because of a past or impending problem. They are then apt to suffer in later times of crisis. Lighthouses have often been built following some marine disaster. The most famous lighthouse of all, the Pharos of Alexandria, was one of the Seven Wonders of the Ancient World, built between BC285 and 247. It was destroyed by earthquakes nearly 1600 years later. Almost at the same time as this was happening, a French sailing ship, the *St Marie de Bayonne*, was driven across the English Channel and onto the rocks along the south coast of the Isle of Wight. The crew were able to abandon ship and, together with the local people, salvaged the cargo, 173 barrels of fine, white wine. They were led in this activity by the local lord, Walter de Goditon. However, the wine belonged to the monks of Livers Monastery in Picardy and de Goditon was therefore committing an offence, not only against the King, who was attempting to maintain a peace with France, but also against the Church of Rome. Pope Clement IV became involved, and after a long and complicated trial, de Goditon was given a fine and ordered by the church, on threat of excommunication, to build a lighthouse on the cliffs, with the light of a fire to warn mariners at sea, and a chapel in which a priest would sing masses for the soul of de Goditon, his family and those at sea. Completed in 1323, the lighthouse was operational until King Henry VIII dissolved the monasteries 200 years later in 1536. Although the chapel of St Catherine's Oratory was destroyed in the dissolution, the lighthouse still remains and, at almost 700 years old, is one of the oldest in existence.[35]

In the same way, monitoring of ecosystems has frequently been instigated following a problem, or if one seems imminent. The observational series from Windermere were begun after the collapse of the lake's perch fishery, the Continuous Plankton Recorder Survey was started to help satisfy the increasing need for fish in the North Sea, the roadside monitoring at Bibury developed out of an investigation of pesticides, and the CalCOFI programme off California was begun because of a collapse of the sardine fishery in the area. Sometimes such programmes have stumbled over the obstacles caused by shifts in funding priorities. There is an increasingly urgent need to sustain these kinds of programmes monitoring biological communities, for they have the potential to illuminate climate changes and their impacts at an early stage. Such warning lights will be an important aid to navigate us through the turbulent conditions of the coming decades that will be caused by global warming.

9

DRUNKEN TREES IN
THE GREENHOUSE

'The climate system is an angry beast and we are poking it with sticks'
(Wallace S. Broecker)

On the northern coast of Britanny, at Porz Guen on Kernic Bay, is a grave whose stones are coated with seaweeds and barnacles. Late on an April day, in heavy rain when its rows of stones were disappearing under the waves, it seemed especially desolate. The tomb, whose shape can only be seen fully at low tide, is a gallery grave, a rectangular chamber with no passage to it. It has stood on this site for 4500 years. When it was first erected, the line of uprights on either side of the gallery had capstones and helped to consolidate a burial mound which faced down into a low valley containing a river. The mound has long since been washed away, leaving on its west side a run of low stones curving round to link with a terminal cell at the higher north-east. This gallery entrance formed a sort of forecourt roughly 6 m square. Looking down under a dark sky and through unbroken rain on this monument lapped by the sea, the magnitude of climate changes in the past, the rises of sea level following the last ice age, are revealed very clearly. What's more, these changes were not caused by people burning fossil fuels or cutting down forests. Can we then be sure that the changing weather patterns we are currently experiencing are no more than some natural fluctuation, perhaps the cobra rising up from the mathematical grass?

Two hundred years ago, during the Little Ice Age, there was more concern about global cooling rather than global warming. In the 1770s, the Lunar Society of Birmingham, whose members were some of the leading thinkers of their day, regularly gathered to discuss their wide-ranging scientific interests.[1] From 1765 until 1809, the Society met every month in members' houses on the Monday nearest the full moon (when there was most light for the journey home), and in its day it was second only to the Royal Society as a gathering place for scientists, inventors, and natural philosophers. Among its members were Matthew Boulton the industrialist, James Watt the developer of the steam engine, Erasmus Darwin, a leading physician who argued for the evolution of life before his grandson Charles, Joseph Priestley the discoverer of oxygen, and William Herschel who discovered the planet Uranus. In addition, the American statesman Benjamin Franklin was a corresponding member of the society, as were others including John Smeaton, the great civil engineer who solved the problem of building a lighthouse on the Eddystone rocks off Plymouth with a design that has been used ever since.[1]

One of the Lunar Society's members was William Small, who, when working in America, had been a science and mathematics teacher to Thomas Jefferson. He had been born in Scotland and believed that its highlands had once supported trees. As they no longer did so, he reasoned that the hills must have become barren and treeless due to a southward spread of a colder climate. Small summarised his view as: 'I am led by this...to suppose, nay to believe, that the frozen space of the Globe is annually increased at the rate of about the 300th part of a degree of latitude...so that after a certain number of years all Europe, and finally the whole surface of this earth, will be frozen, as the Moon is now and has long been.' He even had a scheme to deal with this horrifying prospect, namely to use gunpowder to blow up the polar ice, thus creating icebergs which could be towed to the tropics. There the icebergs would serve as air-conditioning units, rendering the tropics more temperate and habitable.[2]

Concerns over global cooling continued to more recent times. In the late 1960s and early 1970s there was actually a short-term drop in global temperatures, which led to some speculation that this might be the early stages of an ice age![3] The global cooling began even earlier, stopping the German advance on Moscow in the winter of 1941 by freezing the grease in the guns and killing thousands of soldiers. George Kulka, a climatologist from the Czech Academy of Sciences, warned that an ice age was due at any time. Stephen Schneider at NASA's Goddard Space Flight Centre in New York published calculations showing that the cooling effect of rising aerosols would overwhelm the effects of greenhouse gas emissions. However, these few early investigations proved to be incorrect. The science of ice ages and how they relate to planetary wobbles has progressed, and it is no longer thought that another such freezing is imminent. Schneider himself subsequently reported that his previous aerosol calculations were incorrect, being based on inadequate data. It seems likely that the mid-century cooling was caused by the eruptions of a cluster of medium-sized volcanoes pumping sulphate aerosols into the upper air.

Now, the current concern is not with global cooling but with the warming attributable to the increasing concentration of the gas carbon dioxide in the atmosphere. The burning of fossil fuels, such as coal and oil, and natural combustibles such as wood during deforestation have led to a steady increase in carbon dioxide emissions since the Industrial Revolution. Since 1780 this has resulted in a 30% increase in carbon dioxide concentrations in the lower atmosphere. But, if climatologists were so wrong about the global cooling, can we believe their current forecasts of the warming these emissions will produce? The answer to this question is that the earlier studies were based on flimsy data by young researchers starting out, while the present consensus is grounded on vastly more research into how and why climate changes.

The large sensitivity of the surface temperature of the earth to increases in atmospheric carbon dioxide was first realised by the Swedish chemist Svante Arrhenius in 1896, but he was developing ideas first put forward by Jean-Baptiste Fourier and John Tyndall. Fourier was born in 1768 and, after periods of imprisonment when he was in danger of the guillotine, he was appointed in 1798 onto Napoleon's eastern expedition. The French army was accompanied by 150 scientists, called the *corps des savants*, who spent three years roaming the Nile, investigating everything from pyramids and hieroglyphs to crocodiles and scarab beetles, and subsequently revealing the wonders of Egypt to Europe. Fourier was made Governor of Lower Egypt and, while cut off from France by the English fleet, organised the workshops on which the French army had to rely for their munitions of war. His interior thermostat seemed not to readjust after his return from Egypt with the consequence that he never ventured out in any season without an overcoat and a servant bearing another in reserve. Upon his return to France, he became the Prefect of Grenoble, where, perhaps because of problems with body-temperature, he made experiments on the propagation of heat. These he published in his classic treatise, *Théorie Analytique de la Chaleur*.

One day in the 1820s, Fourier began to ponder why the heat from the Sun's rays is not lost after bouncing off the great oceans and landmasses of the world. In 1824, he proposed that the atmosphere behaves like the glass cover of a box exposed to the sun by allowing sunlight to penetrate to the Earth's surface and then retaining much of the 'obscure radiation', now called infrared radiation, which emanates from the Earth's surface. This process has subsequently come to be called 'the greenhouse effect'.[4]

Interestingly, the principal mechanism operating in a greenhouse is not the trapping of infrared radiation, but the restriction of heat loss by convection after the air has been warmed by contact with the ground on which the sun has shone. The trapping of terrestrial infrared radiation is inconsequential, as the American physicist Robert W. Wood demonstrated with an elegant experiment in 1909. He showed that when the glass in a model greenhouse was replaced by rock salt, which is transparent to infrared radiation, it made no difference to the temperature of the air inside. What matters is that the incoming solar radiation enters through the glass, and once the ground and adjacent air is warmed up, the glass prevents turbulent movement of the air carrying the heat away too rapidly. Thus, if the doors of a greenhouse are opened on a cold and windy day, the trapping of heat is much less effective.[5]

Although the trapping of infrared radiation proposed by Fourier may not actually contribute greatly to warming a greenhouse, the mechanism is important in the atmosphere. Fourier's idea was further developed by a self-made man of science who had risen from poor circumstances in Ireland. John Tyndall was a physicist, noted, among other things, for the Tyndall Effect, the diffusion of light by large molecules and dust, which led him to propose that the blue of the sky is due to the scattering of the sun's rays by molecules in the atmosphere. Tyndall's experiments also showed that molecules of water vapour, carbon dioxide and ozone are the best absorbers of heat radiation, and that even in small quantities these gases absorb much more strongly than the rest of the atmosphere. He concluded that among the constituents of the atmosphere, water vapour is the strongest absorber of radiant heat and is therefore the most important gas controlling the Earth's surface air temperature. Six months before Charles Darwin published *On the Origin of Species*, Tyndall demonstrated these experiments before Prince Albert, concluding: 'To the eye, the gas within the tube might be as invisible as the air itself, while to the radiant heat it behaved like a cloud which was almost impossible to penetrate. Thus the bold and beautiful speculation has been made experimental fact. The radiant heat of the sun does certainly pass through the atmosphere to the Earth with greater facility than the radiant heat of the Earth can escape into space.' Tyndall said that without water vapour the Earth's surface would be 'held fast in the iron grip of frost'. He later speculated how changes in water vapour and carbon dioxide could be at the root of ice ages.

Tyndall's strong, picturesque mode of seizing and expressing things gave him an immense influence both in speech and writing throughout the 19th century, and in disseminating a popular knowledge of physical science. Unfortunately, he suffered from sleeplessness throughout his life, and as this became worse with age he experimented more and more with drugs. Finally, in 1893, he died from an overdose of a chloral compound accidentally administered by his wife.[6]

A few years later, in 1896, the Swedish chemist Svante Arrhenius used the recent knowledge of solar spectra and of the spectroscopic absorption bands of carbon dioxide and water to develop the first detailed theory of the Earth's radiation budget. He had begun these calculations while alone in the Nordic winter a couple of years earlier during the breakdown of his marriage.[4,7] Arrhenius did very little research in the fields of climatology and geophysics, and considered his work in these fields a hobby. His major achievement was explaining the electrical conductivity of ionic solutions by presuming that compounds dissociated into oppositely charged ions whose motions constituted the electric current. Arrhenius developed this theory in his 1884 doctoral thesis which, although it passed the thesis defence at only the lowest grade and thereby made him ineligible for university lectureships in Sweden, won him the Nobel Prize for chemistry in 1903. He also discovered the Arrhenius Rate Law, which describes how the rates at which chemical reactions occur change with temperature, and made the suggestion that life on Earth had originated from space-travelling spores, an idea that has many modern adherents.

Arrhenius' intention in his theory of the radiation budget was to apply basic scientific principles to make sense of existing observations, and to go on to develop a theoretical cause of the ice ages. His theory revealed the large sensitivity of surface temperature to increases in atmospheric carbon dioxide. This was partly attributable to the positive feedback due to water vapour (the dominant greenhouse gas), whose atmospheric loading increases with temperature. Allowing for the water vapour in the atmosphere makes the greenhouse effect increase with warming. Arrhenius argued that variations in trace constituents—namely carbon dioxide—of the atmosphere could greatly influence the heat budget of the Earth. Using the best data then available to him (and making the many assumptions and estimates that were necessary), he performed a series of calculations on the temperature effects of increasing and decreasing amounts of carbon dioxide in the Earth's atmosphere. His calculations showed that the 'temperature of the Arctic regions would rise about 8 degrees or 9 degrees Celsius, if the carbonic acid increased to 2.5 to 3 times its present value. In order to get the temperature of the ice age between the 40th and 50th parallels, the carbonic acid in the air should sink to 0.62 to 0.55 of present value (lowering the temperature 4 degrees to 5 degrees Celsius).'[8]

A simple calculation illustrates the important effect of the atmosphere on the planet's temperature. Each square metre of the Earth's surface receives 1353 watts of energy from the sun, of which about 31% is reflected back to space. If the planet's temperature is not increasing or decreasing, all of the energy intercepted by the planet's disc must eventually be radiated back into space by its spherical surface. All objects release energy in the form of electromagnetic waves at a rate that depends on the temperature. The best absorbers usually make the best emitters of radiation, and these are called blackbody radiators. They absorb all the radiation falling on them, that is they do not reflect or transmit it. A blackbody radiator emits energy at a rate (E) per square metre per second which is proportional to the fourth power of the absolute temperature, a rule that is expressed in the Stefan–Boltzmann law:

$$E = \sigma T^4$$

In this equation, T is the absolute temperature as measured in degrees Kelvin (same size as celsius degrees) from absolute zero ($-273°C$). This means that a temperature on the Kelvin scale is obtained from that on the everyday celsius scale by adding 273. σ is a constant of nature. If the Earth can be considered to be a black body and this heat flux (E) equals the 934 watts per square metre of solar radiation that the planet absorbs without reflection, the average surface temperature of the Earth would have to be 255 K or $-18°C$. In reality, the average surface temperature of the planet is well above this at about 288 K or 15°C. The discrepancy of 33° occurs because the temperature calculated is that at which planet radiates to space (the temperature of the upper atmosphere), and not that at the ground. It is the trapping of heat by atmospheric gases, and in particular by water vapour and carbon dioxide, that was described by Fourier, Tyndall, and Arrhenius, which generates this difference in temperature between the surface and the upper atmosphere.[9]

Arrhenius took a year carrying out calculations a modern computer would do in seconds. He was primarily concerned with climatic cooling and considered that any future warming would be comparatively benign and might, on balance, be beneficial to the world. However, he had tackled a problem, the origin of ice ages, that interested no one else at the time, and so his paper was quickly forgotten. During the first few decades of the 20th century, a number of other competing theories were proposed, and serious doubts developed about the importance of changing CO_2 as a factor in the Earth's climate. The idea was revived in 1938 when a quiet and retiring English steam engineer, inventor, and amateur scientist, Guy Stewart Callendar, analysed a set of data on atmospheric CO_2 content taken at Kew near London.

Callendar concluded that at around 1900 the atmosphere over the North Atlantic region contained 274 parts per million (ppm) of CO_2. Then, after arguing that only a small fraction of the CO_2 from the combustion of fossil fuels would dissolve in the ocean, he estimated that the concentration would rise to 290 in 1936, 315 in 2000 and 352 in 2100. Based on a simple model of the absorption of infrared radiation, Callendar concluded that temperature was rising at 0.03°C per decade. His result was based on many assumptions and he used no contemporary CO_2 data on which to base his estimates, but, nonetheless, his prediction was not far from the truth, and his work rejuvenated the CO_2 theory of climate change, so much so that, for a time, it became known as the Callendar effect. After he died in 1964, others took credit for the theory, mainly because his work was based on inadequate observations, but perhaps also because he was not a professional scientist.[4,10]

One scientist who did pick up Arrhenius' and Callendar's ideas was the atmospheric scientist J.S. Sawyer, whose criteria for climatic impacts were described in Chapter 7. In 1972, he wrote a paper drawing attention to the work of Tyndall, Arrhenius, and Callendar, predicting that the 25% increase in carbon dioxide expected to occur by 2000 would lead to an increase of about 0.6°C in world temperature.[11] This was in spite of the observation that global temperatures had been falling up to the 1970s. His prediction of the reversal of this trend proved to be correct and the rise was 0.5°C.[12]

The atmospheric concentration of CO_2 has been monitored closely at Mauna Loa, an observatory on an extinct volcano in Hawaii since 1958 (Fig. 9.1), and, for nearly as long, at other sites including the South Pole. These measurements have been made by the group set up by the late Charles Keeling at Scripps Institution of Oceanography in California.[4] The project was initiated following a disagreement between the young Keeling and his professor, Harrison Brown, at California Institute of Technology. It was Brown's contention that the carbon dioxide in the lower atmosphere is in balance with the CO_2 in the oceans and seas, while Keeling argued that the two might just as likely be out of balance. After gaining Brown's approval, Keeling began making observations to find out, measurements that gradually revealed the steady rise in the concentration of CO_2 in the atmosphere.

Keeling was a meticulous scientist who demanded the same standards from those working with him, characteristics which meant that he was not always an easy colleague to work with.[13] His careful observations have shown that the rise in CO_2 is at a rate consistent with estimates of fossil-fuel emissions and ocean uptake. The data also show an annual rise and fall, which is the annual carbon cycle in the biosphere. During the spring and summer, trees and other vegetation of the northern hemisphere take in CO_2 from the air, thus producing the low points on the curve (the contribution from the southern hemisphere where there is less land is weaker). In the autumn, when nature strips trees of their leaves, decaying matter releases its stored carbon and the curve rises once more. These oscillations are the rhythmic breathing of the planet.

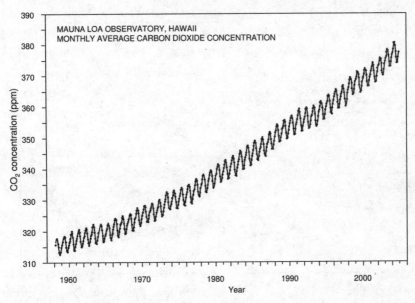

FIG 9.1 CO_2 concentrations measured at Mauna Loa, Hawaii, since 1958, showing the upward trend and seasonal changes.[14]

Measurements of the concentrations of ancient gas trapped in Antarctic ice cores show that the steady rise observed by Keeling's team is only a comparatively recent phenomenon (Fig. 9.2). Snow deposited at the polar regions is gradually compacted as further snow falls, becoming solid ice and trapping within it tiny bubbles of air. Drilling out an ice core through such ice and examining the composition of the air in the bubbles allows scientists to determine the atmospheric content as a function of time, as samples at the top are more recent than those at the bottom. Over the millennium of the pre-industrial period, CO_2 levels were constant at about 280 parts per million by volume (ppmv), before rising steadily, after the dawn of the industrial revolution in about 1750, to nearly 390 ppmv, an increase of 37 per cent. The rise generally matches the ongoing increase in CO_2 emissions over the same interval. Despite efforts to curb carbon dioxide emissions, global output has grown four times as fast since 2000 as during the previous decade.[15]

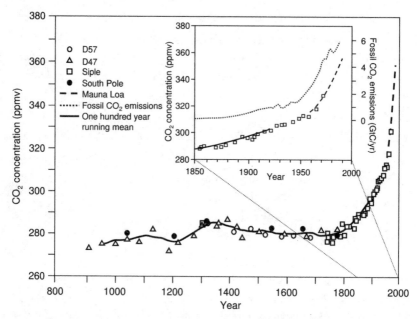

FIG 9.2 CO_2 concentrations over the past 1000 years from ice core records in Antarctica and (since 1958) in Mauna Loa, Hawaii. The rapid increase in CO_2 concentration since the onset of industrialisation is evident and has followed closely the increase in CO_2 emissions from fossil fuels (see inset period from 1850 onwards).[14]

Alongside this increase of CO_2, there have been similar changes in the concentrations of other greenhouse gases during the industrialised period: concentrations of methane (a gas 21 times more potent than carbon dioxide) have more than doubled since the mid 19th century, while nitrous oxide concentrations have increased by 8%. The introduction of chlorofluorocarbons (CFCs) into the atmosphere has also provided a new source of greenhouse gases. The overall effect of the build-up of all these gases is equivalent to increasing the solar radiation being trapped at the Earth's surface by 2.5 watts per square metre between 1850 and the present day, about 1% of the average amount of energy radiated by the Earth back to space. CO_2 has contributed some 60% of this figure, methane about 25%, and nitrous oxide and CFCs the remainder.

It is sometimes claimed that volcanoes emit more CO_2 than human activity, and so could easily be responsible for the increase in greenhouse gases. This is not true: total emissions from volcanoes on land, at 0.3 gigatons of CO_2 each year, is only about a hundredth of human emissions (a gigaton is one billion tons). However, CO_2 emissions due to human activity (26 gigatons per year) are small compared to other natural sources such as biological processes and erosion of chalk deposits (440 gigatons per year from the land). This might seem to imply that industrial contributions to the atmosphere will not significantly affect the world's atmosphere, but this is not the case, for the natural sources are balanced by equally large natural sinks: the photosynthesis of terrestrial plants and absorption by the ocean. Human emissions have now disturbed this balance. Ice cores show that CO_2 levels in the atmosphere remained fairly steady at between 180 and 300 parts per million for the past half-million years, only to shoot up to more than 380 ppm since the industrial age began. These changes over the past 1000 years are shown in Fig. 9.2.

We can be confident that people are responsible for the recent extra CO_2. Fossil fuels contain virtually no carbon-14. This is an unstable isotope of carbon formed when cosmic rays hit the atmosphere. It decays with a half-life of 6000 years, which means that half of the carbon-14 atoms present disappear every 6000 years. Over the millions of years that fossil fuels have been waiting to be dug out, any carbon-14 atoms will have long decayed, so that when the carbon is burned the resulting CO_2 will also contain no carbon-14. Samples of the carbon contained in tree rings record the atmospheric conditions and can be dated by counting the rings. Studies of tree rings in this way have shown that during the industrial age, from 1850 to 1954, the carbon-14 in the air dropped by 2 per cent, a value roughly consistent with the increase in the CO_2 content of the atmosphere.[16] Unfortunately, after 1954 nuclear tests released large amounts of carbon-14 into the atmosphere, and this masks any signal from fossil fuels.

Further warnings that the recent rise in CO_2 levels will usher in a warmer world come from observations of how climate has varied over several tens of thousands of years, data which reveal the greenhouse

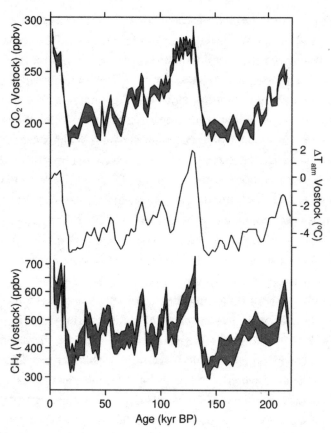

FIG 9.3 A comparison between carbon dioxide and methane concentrations in an ice core obtained at Vostok, Antarctica, and the estimated changes in temperature over the past 220,000 years.[14]

effect in operation. The results from one such core drawn from the Russian Vostok Station in Antarctica in Fig. 9.3 show a strong correlation between the global surface temperature and the concentrations of carbon dioxide and methane. As the concentration of these gases has increased or decreased, global surface temperatures have gone up or down in line with the predictions of Arrhenius, Callendar, and Sawyer. It is important to note that this graph does not show the present carbon dioxide concentration: at over 380 ppmv, it is way above the top of the box!

A criticism that has been levelled against the ice-core data from Antarctica and Greenland is that the level of CO_2 in the atmosphere often did not appear to rise until temperatures had been climbing for some time. The delay might sometimes have been about 800 years, but there is uncertainty about the precise timing, in part because the air trapped in the cores is sometimes younger than the ice itself. While fairly basic physics demonstrates that greenhouse gases such as CO_2 will trap the heat radiating from the earth and so make the planet warmer as their presence increases, this does not mean that there will be a perfect match between past temperatures and past CO_2 levels. Many other factors, such as periodic variations in the earth's orbit and major volcanic eruptions, also affect the temperatures and CO_2 concentrations, and these can easily obscure the relationship between CO_2 and temperature.[15]

During one of the fluctuations in Fig. 9.3, the Eemian interglacial 125,000 years ago, temperatures may have been 1 to 2°C warmer than today, with sea levels 5 to 8 metres higher than at present. It is therefore tempting, as some have done, to dismiss the current warming as something that has happened before as a result of purely natural processes. This would be a mistake. The natural factors operating at this earlier time, such as differences in the amount of solar radiation reaching the Earth's surface, can account for, at most, a small part of the recent warming, so the present changes in climate are not mainly attributable to the causes operating in the past. Nor is it safe to say that the future warming is nothing to worry about; the earlier rises in sea level would have been large enough to submerge many major cities.

Much scientific effort is now being employed in setting up sophisticated computer models to determine the future climatic impacts of the continuing increase in greenhouse gas emissions, not just globally, but on the regional scales that are of more concern to the local populations and national politicians. In spite of the limited abilities of the general circulation models of the atmosphere and oceans to reproduce all the details of the current global circulation patterns, they are the only realistic way to predict the impact of human activities on the climate. As these models are very complex and the number of processes that need to be considered is extremely large, only a few research centres in the world can undertake such modelling. Most of the models are based on those used for daily weather prediction, but in addition, since climate change acts over very much longer timescales than the weather, the interactions of the atmosphere with the oceans, land, and ice-cover, as well as effects on the biosphere, also have to be included.

Incorporating detailed processes is not straightforward, and will not necessarily reduce the uncertainties of the predictions. When scientists at the UK Hadley Centre were setting up their latest climate model HadGM2-ES they put too little plant life in the world's arid regions, leaving insufficient vegetation to hold the soil in place. Soon dust levels in the model atmosphere reached three times higher than average, which in turn fertilized the ocean causing phytoplankton blooms. The unrealistic dust storms revealed that even a small amount of plant life in the world's deserts can influence the whole climate.

Even so, although different models predict a considerable range of future climatic conditions, they are still consistent in their broad prediction that the build-up of greenhouse gases in the atmosphere is beginning to exert a dominant influence on the global climate. It is these results that have been distilled into a consensus by the Intergovernmental Panel on Climate Change (IPCC), a unique and unprecedented statement by a global community of some 3500 scientists. The models agree that the warming is expected to be considerably greater at high latitudes, especially in the northern hemisphere, than in the tropics, and this is consistent with the generally held view of past climate changes, notably the ice ages, which is that the tropics have been more stable than higher latitudes.[16]

Observations are now starting to show the rise in global temperature expected from the increases in the concentrations of greenhouse gases. A graph of the Earth's mean surface temperature from 1854 to the present (expressed in Fig. 9.4 as a deviation from the average temperature 1961 to 1990) reveals a clear rise over the past century. The annual surface temperatures have risen by about 0.6°C, a rise that has been slightly exceeded

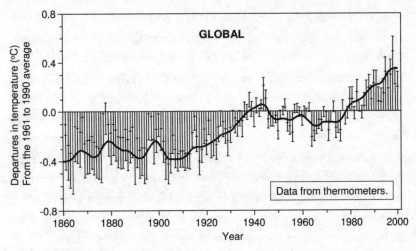

FIG 9.4 Combined global annual land-surface, air, and sea-surface temperatures (°C) from 1861 to the 1990s relative to 1961–1990.[14]

in the southern hemisphere. In addition, the eleven warmest years in the record have all happened since the 1990s. But this evidence on its own is not conclusive proof that the global warming has been caused by the emissions of greenhouse gases. For example, at the start of this record in the mid 1800s, Europe was experiencing the cool period called the Little Ice Age, which began in the 16th century and continued to the reign of Queen Victoria. Figure 9.4 could merely be showing a period of warming following a prolonged cooling event.

A few scientists have gone even further and suggested that, even in spite of the evidence for rising levels of CO_2 in the atmosphere, and predictions from basic physics of what it is likely to do, the recent trend in global temperature is part of an oscillation whose cause is not on Earth, but in the Sun. This theory has its origin in the observation that the number of sunspots on the surface of the sun seem to have increased during the past century in the same way as have global temperatures. Coincidentally, the Little Ice Age in the 17th and 18th centuries, when European rivers such as the Thames intermittently froze sufficiently to support festivities, corresponded with a period during which there were almost no sunspots, an interval known as the Maunder Minimum. The sun, of course, exerts a crucial influence on the Earth's climate and the total amount of energy reaching the Earth varies, but, even so, the observed variations in its heat flux are not large enough to explain the recent warming. In the late 1990s, Henrik Svensmark and colleagues at the Danish National Space Centre suggested that the sun might influence the world's weather by its action on cosmic rays. They conjectured that cosmic rays affect cloud formation by ionising the atmosphere and, although most cosmic rays come from deep space, changes in solar activity will alter the number of cosmic rays reaching the Earth. When there are many sunspots, the sun's magnetic field strengthens, which deflects more cosmic rays from the Earth. The Danish scientists suggested that fewer cosmic rays would mean fewer clouds and a warmer planet.[17]

Unfortunately, much as we might want to shift the blame for rising temperatures to a distant natural process, we are not able to do so: the case against the sun and cosmic rays does not stand up to close inspection. By far the dominant variation in the numbers of sunspots is the cycle of maxima and minima that recur about every 11 years. There is no convincing evidence for this cycle appearing at all clearly in the global temperature time-series. The trend in sunspot numbers which follows an upward curve like the temperatures is the considerably smaller increase in the peaks of the 11-year cycle from one cycle to the next. Why should this weak fluctuation affect the climate while the much more striking 11-year signal does not do so? Further, the association of the Little Ice Age with the Maunder Minimum also does not withstand examination in closer detail. Certainly, as already described in Chapter 2, there were low temperatures in Europe and the North Atlantic during this period, but this seems to have been a local phenomenon, for there appears to have been little drop in the average temperature of the northern hemisphere at this time[18,19] (Fig. 9.5).

A record of temperature changes based on actual measurements is only possible from some time after thermometers became widely used in about the early 1800s, which is why the graph in Fig. 9.4 begins in 1854. To get temperatures further back in time it is necessary to infer them from proxy measurements such as the growth-rings in trees and similar measurements from corals. The thickness in each year's ring shows how much growth was made by the tree or coral, and hence how good the weather was. Extending the temperature series back across the past 1000 years in this way was first attempted in 1999 by Michael Mann and his colleagues at the University of Virginia.[20] There are considerable differences in the reconstructions that have been produced in this way, and this is why the uncertainties shown in the graph are much wider in the period before temperature measurements were made. Even so, different reconstructions generally agree over two aspects: the relative warmth during the early middle ages from 1000 to about 1300, and, most strikingly, the sharp upturn in temperatures at the end of the 20th century. This rise reaches values much higher than at any earlier time, and has led to the graph being referred to as the 'hockey-stick graph'.

The graph shows there was no pronounced drop in temperature associated with the Little Ice Age period from 1600 to 1800; while some of the reconstructions do show it, in others it is absent. In addition, it is not really seen in data from the southern hemisphere. This then raises the question of how would it be possible for a fluctuation in cosmic rays to cool only one region of the Earth in the Little Ice Age and yet cause the recent rise in global temperature?

There are also difficulties with the proposed mechanism of cosmic rays triggering clouds and so affecting temperatures. It hypothesises that the ionisation of the air by cosmic rays imparts a charge to aerosols, encouraging them to clump together until they cause the condensation of water and the formation of clouds. As yet, there is no convincing evidence that such clumping occurs. In any case, the atmosphere already has plenty of such cloud condensation nuclei, so it is not clear that cosmic rays will have any great effect on cloud formation. The mechanism also requires that the extra heat that clear skies allow in during the day outweighs the increased heat losses at night. This is not necessarily so, as the physics of

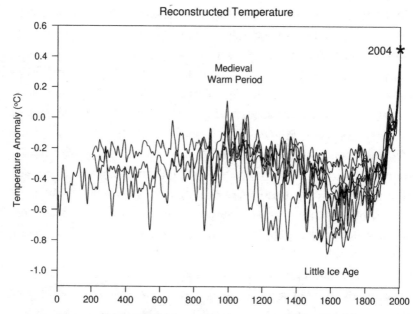

FIG 9.5 The spread of northern hemisphere temperatures for the past 1000 years from different reconstructions. The part from the 17th century onwards is based on thermometer measurements.[14]

clouds is more complicated. Finally, the mechanism requires a downward trend in the flux of cosmic rays to drive the recent rise in temperatures but direct measurements of the cosmic ray flux going back 50 years show only a periodic fluctuation in intensity without any steady decline.[17]

In summary then we cannot put the blame for rising temperatures on the sun. Further, Fig. 9.5 shows that the magnitude of the most recent warming is much larger than any seen in the last 1000 years, and so natural fluctuations in the world's weather system are also an unlikely cause. Given that it is expected on simple physical grounds that the greenhouse gases we are releasing into the atmosphere will have this effect, there would seem to be no need for another explanation. Even so, supporting evidence for global warming occurring is not easy to obtain, since there are many natural phenomena which can confuse or mask the signal. Large volcanic eruptions are one such process. These can significantly affect the subsequent weather around the world because of the cooling caused by large amounts of ash and sulphur dioxide in the upper atmosphere. Dust from the Mt Pinatubo eruption in the Philippines during June 1991 caused spectacular sunsets all around the world for many months after the eruption, produced a 2% reduction in the amount of solar radiation reaching the Earth's surface and lowered the global average temperature by 0.25°C during the following two years.

Sometimes the breaking of climatic records, e.g. the driest summer or the wettest January, is cited as evidence of climatic change. However, climate extremes are not, in fact, unusual and every month somewhere in the world a climate record is broken. This is not necessarily evidence that our climate is changing, but may show only that our data records are limited and that global climate patterns are complex. In the search for evidence, it is necessary to look at the records from many sites and over long periods, and not be misled by what turn out to be short-term variations. Thus, although the period since the late 1980s has been the warmest since accurate measurements began a century ago, and although the numbers of hurricanes, gales, and droughts seem to have become higher, it is still not possible to conclude that this is clinching evidence for global warming.

Nevertheless, there are other observations that support the hockey-stick temperature data, and, in agreement with expectations based on warming caused by greenhouse-gas emissions, these come from close to the poles. About ten years ago, at an international conference, a professor from the former Soviet Union told me over lunch that a striking feature of the far north of his country was the way trees were beginning to lean over. We were enjoying the summer in the very different conditions of Avignon in the south of France. The northern trees had begun tilting because in many areas of both Alaska and Siberia the permafrost, the frozen ground which provides a solid foundation for much of these northern regions, has warmed to within one degree Celsius of thawing. Vladimir Romanovsky, who studies permafrost at the Geophysical Institute of the University of Alaska, has obtained a record of the thawing along a 1200-mile transect in Siberia and compared it to data from Alaska. Warming of the permafrost in the two areas was quite similar. In Fairbanks, Alaska, and in the eastern Siberian city of Yakutsk, for example, the permafrost has warmed about 1.5°C during the past 30 years. Over large areas of the far north, forests appear to be sinking or drowning as melting ice forces water up. Alaskans have taken to calling the phenomenon 'drunken trees'. Hydraulic jacks are increasingly being used to keep houses from slouching and buckling on foundations that used to be solidly frozen all year.[21] In Barrow, the northernmost city in North America, people now have to cope with mosquitoes in an area where they were once rare, and to rescue hunters trapped on breakaway ice at a time of the year when such things were once unheard of. A few hours' drive from Anchorage, beetles have killed a four-million-acre spruce forest, the largest loss of trees to insects ever recorded in North America. Government scientists have estimated that rising temperatures allowed the beetles to reproduce at twice their normal rate.

This warming in the vicinity of both the North and South Poles has revealed itself most dramatically in the rapid disappearance of snow and ice. Back in 1997, William de la Mare, a researcher at the Australian Department of the Environment, reported that the ring of sea ice around Antarctica had retreated abruptly, between the mid 1950s and the 1970s by nearly 3 degrees of latitude.[22] This change occurred before the first satellite capable of monitoring sea ice was launched in 1973; it was observed in the records of sea ice which were kept by whaling ships.

Whaling began in the Antarctic in 1904, and the first factory ships arrived in 1905. The whalers quickly discovered that whales tend to concentrate around the ice edge and so, as they recorded the position of each catch, they were also recording the extent of the ice. Between the 1950s and the 1970s, the whalers had been forced to extend their hunt progressively farther south as the animals followed the retreating ice edge. In pursuing their quarry into inhospitable regions, they gleaned almost the only information about these areas available at the time. De la Mare retrieved these records from the International Whaling Commission in

Norway and extracted from them the southernmost catches for each year at each longitude. His data showed that the summertime edge of the sea ice had moved southward from 61.5 degrees south, on average, to 64.3 degrees, a 330 km southward shift which represents a 25% reduction in the area of sea ice.

These climatologically valuable records are almost all that is left of Antarctic whaling. By the late 1950s, the whalers had nearly wiped out the stocks of blue, humpback, and fin whales in the Southern Ocean. They then turned their attention to sei, and finally to minke whales, until commercial whaling was banned altogether in 1987. Now the most important fishery in the Southern Ocean is for krill, the creatures the whales used to eat.

Satellite measurements have extended monitoring of polar-regions beyond the 1970s. Observations over the period 1978 to 2005 show that the area of ice in the northern hemisphere at the end of summer has decreased at a rate of 8% per decade (Fig. 9.6), a rate that seems to have been accelerating since, 2002. In previous centuries, explorers struggled unsuccessfully to find a north-west passage around the north of Canada and a north-east passage around the north of Siberia. Perhaps the most famous of these early expeditions was that of Sir John Franklin, whose entire complement of Sir John and 28 men died after their ships became icebound near King William Island in the Canadian Arctic. These passages are now sometimes open to shipping at certain times of the year. Julienne Stroeve of the US National Snow and Ice Data Centre has forecast that, if current rates of decline in sea ice continue, the summertime Arctic could be completely ice free well before the end of this century.

What ice remains is also getting thinner. Measurements of thickness have been made since the nuclear submarine USS Nautilus did so when it surfaced at the North Pole in 1967. Subsequent data indicate that there may have been an approximate 40% decline in Arctic sea ice thickness in late summer to early autumn between the 1960s and the mid 1990s, with a substantially smaller decline in winter. However, the relatively short record length and incomplete sampling mean that weather variability could have influenced these observations. More recent data have revealed

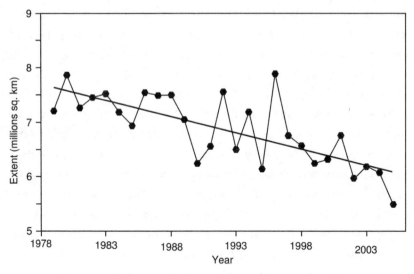

FIG 9.6 Extent of polar sea ice from the end of September 1978 to 2005 (Data from the US National Snow and Ice Data Centre).

more pronounced reductions. In May 2009, Ronald Kwok from the Jet Propulsion Laboratory in Pasadena reported that multi-year ice close to the North Pole had thinned by about 0.7 m between 2004 and 2008.

It is not just ice on the sea that has been shrinking; observations from satellites have shown that the extent of snow cover in the northern hemisphere decreased annually by 1–3% per decade between 1978 and 2005, with greater decreases of 3–5% during spring and summer. One region where the snow appears to be diminishing particularly rapidly is the western United States, especially in the spring. The overall trend has correlated closely with changes in the average temperature at high latitudes. Trends in the southern hemisphere have not been so pronounced because the Antarctic snow cover is permanent, and there is little winter snow cover in South America, Australia, and New Zealand.

John Tyndall's interest in the greenhouse effect was triggered in part by the great glaciers of the Alps and evidence they provided of previous ice ages. One glacier particularly captivated him: the Mer de Glace, which spills into the Chamonix Valley near the hamlet of Les Bois. This glacier has since receded by more than a kilometre, so much so that it is no longer visible from the valley below as it would have been in Tyndall's day.[23] Mer de Glace is but one of many. The International Panel on Climate Change has said, 'The nearly worldwide decrease in mountain glaciers extent and ice mass is consistent with worldwide surface temperature increases'. (Hopefully the error in the rate of melt of Himalayan glaciers in the IPCC report will not continue to dog this discussion.) The Greenland ice sheet is releasing a cubic kilometre of water into the Atlantic as icebergs every two days. After allowing for replenishment by snowfall, the annual loss is 220 cubic kilometres, twice what it was a decade earlier. Virtually all of the glaciers south of the Arctic Circle have speeded up as a result of the 3°C warming of the region. Much of the acceleration is caused by meltwater penetrating crevasses and allowing the ice to slide along on a river of water.[24]

In Antarctica, things are even more alarming. The Antarctic ice cap contains 70% of the fresh water and 90% of the ice on the planet. In February 2002, the Larsen B ice shelf, with an area of the size of Luxembourg, broke up in a matter of weeks.[25] This ice shelf had acted like a dam, slowing the flow of glacial ice into the sea, and almost immediately the glaciers began to flow more rapidly. There have been reports that 90% of the 244 glaciers on the Antarctic Peninsular and 60% of west Antarctic glaciers are losing mass. The west Antarctic ice loss is already contributing 10% to the global rise in sea level.[26]

The longest instrumental records of local sea level (two or three centuries at most) come from tide gauges used to monitor the regular rise and fall of the sea. Based on these, the rate of global mean sea level rise during the 20th century was in the range 1–2 mm/yr. The average rate of sea level rise is larger than during the 19th century, but no significant acceleration in the rate of sea level rise in the 20th century has been detected as yet. Thermal expansion of the oceans as they warm up is a major contributor to this rise in sea level: Sydney Levitus and colleagues in the USA have shown that the heat content of the world's ocean has increased along with the rise in global temperature.[27] Melting of sea ice floating in the

Arctic Ocean and around Antarctica will not contribute to a rise of the sea any more than melting of an ice cube in a drink makes it flow over the top of the glass. The ice is already displacing the seawater in the way it will when melted. (There is a difference because the density of the ice is not the same as that of the meltwater, but this difference is small). But this cannot be said of land-based ice: the calving of glaciers and melting of continental ice sheets have contributed a substantial part to the swelling of the oceans. The Greenland icecap alone contains enough frozen water to raise the world's sea level by 7 m in the future.

Associated with the temperature increase in the ocean, there seems to have been an increase in evaporation. A recent analysis by Ruth Curry at Woods Hole Oceanographic Institution on Cape Cod, Massachusetts, and others, has shown that the salinity of the North Atlantic Ocean has increased, especially during the past decade, and they attribute this rise in saltiness to an extra loss of water at the ocean's surface.[28] If it continues, this evaporation will be accompanied by increased rainfall down stream over Europe.

An important concern arising from the warming at high latitudes is over the impact it will have on the thermohaline circulation in the ocean. If reduced cooling and more abundant freshwater in northern latitudes hinders the sinking of dense water to the ocean abyss, the northward flow of water in the Atlantic Ocean could stall. In 2005, a team led by Harry Bryden, at the University of Southampton, reported that a comparison of observations made in 2004 with similar data from 1957, 1981, 1992, and 1998 showed that a 30% decrease in the northward flow had occurred at 26.5°N.[29] Although they cautioned that this change was based on scarce observations, it still fuelled fears that the Ocean Conveyor Belt could collapse with a consequential slowing of the Gulf Stream and lowering of northern temperatures. However, subsequent analysis by Stuart Cunningham at the National Oceanography Centre in Southampton and his colleagues of 12 months of data from the array of moored instruments deployed during 2004 showed that the observed changes were due to short-term variability and were not the result of global warming.

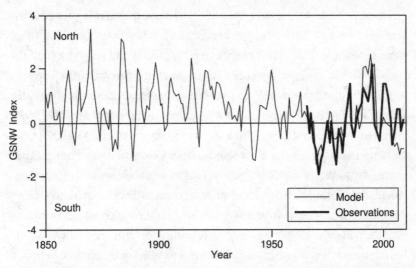

FIG 9.7 Latitude of the Gulf Stream from 1850 to the present from the BRS model. Observed positions are superimposed from 1966 onwards.

The strength of the overturning circulation varied by up to a factor of 8 between March 2004 and March 2005, and Bryden's group had unknowingly probed the ocean during a period of relatively low circulation. Jochem Marotzke from the Max Planck Institute in Hamburg has reported that available data provide no statistically significant evidence for a change in the overturning circulation over the past 50 years.[30]

There are no direct observations of the ocean circulation earlier than the past half century that can be compared with those now being made. Only proxy data can be used to look for changes over longer timescales. The Gulf Stream north wall index is one example. Over the period from the beginning of measurements in 1966, the Gulf Stream has vacillated north and south with no discernable overall trend. If it is assumed that the position in the past has been related to the wind patterns as it has been over the most recent decades, the simple model of Chapter 2 (Fig. 2.5) can be used to hindcast the positions. Again no smooth trend is apparent, but the Stream does seem to have spent slightly more time to the south during the past half century (Fig. 9.7).

There is now little doubt that the greenhouse warming forecast by Fourier, Tyndall, Arrhenius, Callendar, and Sawyer has begun to happen, a conclusion that has been agreed by the world's scientists of the IPCC. Further, the first casualties of the warming are starting to appear. In 1767, Philip Carteret was on a voyage of exploration across the Pacific Ocean for the Royal Navy in an old, poorly equipped and provisioned ship. His first discovery was Pitcairn Island, where, 30 years later, Fletcher Christian and the mutineers from the *Bounty* were to hide. They escaped notice for 19 years, partly because Carteret, who had no chronometer to calculate longitude, had marked the island on his map incorrectly. Then, on the night of 24th August, Carteret discovered a small ring of coral islands, which he called the Carteret Islands. Unfortunately, when he and his crew returned home, these discoveries were eclipsed by the celebrated voyage of James Cook, and Carteret was laid off on half pay.

After more than two centuries, his name is again being remembered, but only because that low-lying group of islands is about to be flooded by rising sea levels. One island has already been cut in half, and the seawater has made soils useless for growing crops. The 1500 islanders are having to pack up and leave. Within a decade, the islands will be under the ocean and Carteret's name will again sink into obscurity with them.[31]

However, the situation is more complex than this, and many islands in the Pacific Ocean are standing up to sea level rise against the odds. Paul Kench at the University of Aukland in New Zealand, and Arthur Webb at the South Pacific Applied Geoscience Commission in Fiji, used historical aerial photographs and high-resolution satellite images to study the changes in the land surface of 27 Pacific islands over the past 60 years. They found that just four islands have diminished in size since the 1950s, and the area of the remaining 23 has either stayed the same or grown. Kench and Webb say that the trend is explained by the islands' composition. They are made up of coral debris eroded from the reefs that typically encircle the islands, which is pushed up onto the land by winds, waves, and currents. Because corals are alive, they provide a continuous supply of material. Causeways and other structures linking islands can boost growth by trapping sediment that would otherwise get lost to the ocean. However, Kench and Webb warn that while the islands are coping now, any acceleration in the rate of sea level rise could overtake the sediment buildup.[32]

From a range of computer model runs carried out in different countries, the way that the world's climate will develop in the future is now clear. The sea level will continue to rise, as will global temperatures, especially those close to the poles. Snow and ice will continue to disappear. How fast these changes occur depends on what happens to greenhouse gas emissions, but also on poorly known details of the processes involved, such as the dynamics of Antarctic glaciers or the way the Ocean Conveyor Belt operates. But these are only the global-scale changes. People and ecosystems will also be strongly affected depending on how these changes vary from place to place, and how they impact on communities.

For instance, Robin Clark and his colleagues at the UK's Hadley Centre have found that regional changes in extreme summer temperature could exceed average global warming by several degrees.[33] They ran 224 simulations of climate responses for an atmospheric carbon dioxide level double that of today's. A fifth of these simulations produced an average global warming of 2°C, and, in these, single-day extreme temperature increases of 6°C or more sometimes occurred in large parts of Europe, North America, or Asia. The regional changes in heatwaves were related to variability in reductions in soil moisture. However, fine details like this are also dependent on the manner in which the global changes in climate modify the large-scale seesaws in the planet's weather, and on the physical and ecological mechanisms we have considered in earlier chapters.

10

AS THE TEMPO OF
THE DANCE HEATS UP

'Prediction is very difficult, especially if it's about the future'
(Neils Bohr, physici st)

John Webster (also known as Johannes Hyphastes, 1610–1682) was an English clergyman, physician and chemist, with interests in astrology and the occult. His chemical book, *Metallographia* (1671), which attributed to minerals the property of growth, was used by Isaac Newton in his alchemical work. The 17th century was a time when there were many accusations of witchcraft, often made by leaders of the church with the victims facing subsequent trials and execution. Perhaps the most infamous of these were the Salem witch trials in Massachusetts during 1692, which led to 31 convictions and 8 hangings. Webster was a severe sceptic of witchcraft, even going so far as to suggest that the Bible had been mistranslated to support the belief. In his book, *The Displaying of Supposed Witchcraft* (1677), he gives some very sound advice for approaching such matters:

…there is no greater folly than to be very inquisitive and laborious to found out the causes of such a phenomenon as never had an existence, and therefore men ought to be cautious and to be fully assured of the truth of the effect before they venture to explicate the cause.[1]

While Webster's warning was vital for the protection of the victims of witchcraft trials, his words have an even wider application to scientific investigations, for scientists have to be sure that the phenomena they are examining really do exist and are not just a misinterpretation of observations or the result of some fluke occurrence. They have to be sure that they are not looking at a Black Swan event. (This term for a high-impact, hard-to-predict, rare event goes back to the 17th century European assumption that all swans are white—the black swans of Western Australia were not discovered until the next century.)

Robert Wood, whose investigation of the physics of greenhouses was described in the last chapter, very much took the same view as was exemplified by his examination of N-rays. Shortly after the discovery of X-rays, the distinguished physicist René-Prosper Blondlot at the University of Nancy announced the discovery of N-rays, which he reported as emanating from most substances, including the human body, with the exception of green wood and some metals. The rays were detected by a screen on which was a suitably prepared deposit of calcium sulfide that glowed slightly in the dark when the rays were refracted through a 60-degree angle prism of aluminum. Observing these rays necessitated looking carefully at a phosphorescent screen in a darkened room. Many scientists and scientific papers claimed to have replicated Blondlot's experiments, but other notable physicists of the day were not able to do so.

Following his own failure and 'wasting a whole morning', Wood decided to visit Blondlot's laboratory and see the experiments for himself. Wood suspected that the French physicist was deluding himself and so, in the darkened room, he secretly removed the prism from the N-ray detection device. Without this prism the apparatus could not work, yet the experimenters still said they were observing the rays. For the next experiment he replaced the prism. Blondlot's assistant saw him tampering with it but thought he had removed it instead. When the experiment was now tried, the assistant said he could not see any N-rays, even though the device was now in full working order. Wood published his findings in Nature,[2] and by 1905 no one outside Nancy believed in N-rays. This incident is a cautionary tale about the dangers of self-deception.

Webster's warning has to be borne in mind when considering climatic phenomena: these can be difficult to pin down because they are often transitory or intermittent, and they can be hard to separate from the noise of daily weather fluctuations. The association of the position of the Gulf Stream with the rises and falls of biological populations in the seas and lakes around Europe, a connection between two highly variable systems operating on very different timescales, is just such a phenomenon. Evidence presented in Figs 1.1, 1.3, 1.4, 3.1 and 3.2 shows that this relationship has been observed in different places, with different species, and within different communities.

It may also have been happening over a longer time than the data in these graphs cover. Although there are no observations of the position of the Gulf Stream's north wall that can be used to calculate the GSNW index before 1966, the model used in Fig. 2.5 can hindcast what they might have shown. In this way, the graphs can be taken back to the beginning of the plankton series in 1948. The resulting data (Fig. 10.1) show lines tracking each other for almost the last half-century. This association does not quite extend to the present day, however, because, in the 1990s, the link abruptly broke as the graphs diverged.[3] A similar sudden breakdown in the relationship has also been observed in the north-east Atlantic,[4] where it occurred a decade earlier without the linkage ever being re-established. Chris Reid at SAHFOS, UK, has called sharp changes in an ecosystem such as these 'regime shifts'. In the case of Fig. 10.1, any regime shift seems to be confined to the central North Sea, because the earlier graphs in Fig. 1.3 show no signs of it extending further north.

Therefore, like some other climatic phenomena, the linkage between the plankton and the Gulf Stream may be of restricted duration, only operating for a few decades, and this transitory nature also appeared in the ERSEM model simulations described in Chapter 8.[5] Even though the

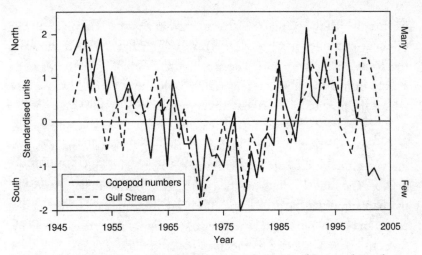

FIG 10.1 Annual changes of zooplankton populations in the central North Sea compared with the position of the Gulf Stream.

phenomenon may be at best episodic, its widespread appearance still justifies venturing 'to explicate the cause'. The association may have only happened over one period of a few decades, but there must be some explanation of how it arose. For a span of 30–50 years, there must have been a fleeting similarity between the weather patterns driving the Gulf Stream and those inflicted upon the plankton. The only other possible drivers of the plankton changes are the change in fish stocks accompanying over-fishing or pollution. Neither over-fishing nor the pollution in the North Sea has been connected with the distant ocean currents, and neither process is likely to have generated the kind of oscillatory behaviour observed.

While this phenomenon does justify the search for a cause, finding the origin of the association is less easy. It seems almost certain that the linkage arises out of ocean-wide patterns in the weather systems but, if so, there is a difference in timescales to be overcome. The Gulf Stream lumbers back and forth about two years after the changes in the winds, while the zooplankton live their hectic lives over a few weeks. A possible way that the link may be made is if the ocean circulation generates the weather which the plankton suffer or enjoy in any year. However, all the evidence examined in Chapter 7 seems to show that any such effects of the ocean currents are buried deep within the variability of the weather arising from atmospheric events well beyond the region. Chapter 8 discussed the possibility of ecosystems being sensitive to these kinds of climatic fluctuations originating in the ocean. In the case of Windermere's water fleas, the sensitivity might arise because of the occasional mistiming when the food of the zooplankton has come and gone before they need it. The model results for the North Sea plankton indicate another mechanism in which the oceanic effects are imprinted on winds, cloud cover, and temperature, and these separate components are extracted by the responses of different members of the community.[5]

The Gulf Stream plankton phenomenon is intertwined with another, the NAO. In many ways, the Gulf Stream north wall index is no more than a manifestation of the NAO, for its shifts appear to be largely a delayed response to how the NAO has varied. As one of the global seesaws which span the world's oceans, the NAO would be acceptable to Webster. There is no doubting the reality of the most gigantic of these oscillations, ENSO. Its effects across the planet are unmistakeable and most of its dynamics have been worked out, even though the way cycles are triggered is still unclear. Even though the case for the NAO is not this strong, its presence in the world's climate system cannot be denied. Much of what has happened to biological communities around the North Atlantic over recent decades can be put down to the effects of the NAO or related to the GSNW index. All the graphs displayed previously (Figs 1.1, 1.3, 1.4, 3.1, 3.2, 6.1, 6.2, and 10.1) show populations from the various ecosystems fluctuating up and down, rather than just drifting steadily in one direction. These oscillations are to a large extent attributable to the atmospheric seesaws. How the processes operating in these phenomena develop in the future will therefore be an important feature of the climate changes to come.

Global warming is certainly a phenomenon whose existence cannot be doubted, and one whose cause has been 'explicated'. The Intergovernmental

Panel on Climate Change has declared that the recent warming is very likely to have been due to human activities releasing greenhouse gases into the atmosphere. Such a source clearly carries much more weight than any individual or group. According to Susan Solomon who headed the US group at IPCC, the report 'is a very rigorous statistical analysis, comparing models in space and time in a more detailed way than ever before.' Two agreements between observations and model predictions have been particularly key: the greater warming over land masses compared with the oceans, and the combination of warming in the lower atmosphere with cooling in the stratosphere. However, in order to make a cast-iron case for the risks of global warming, the IPCC adopted a review process so rigorous that research deemed controversial, not fully quantified or not yet incorporated into climate models, was excluded. This leaves sceptics little room for manoeuvre, but means that many processes not included in current models of the climate, and which could accelerate global warming or sea level rise, were not considered. Among these are the collapse of the Greenland ice sheet, rapid melting in Antarctica, shutting down of the Gulf Stream, and release of carbon dioxide and methane from soil, the ocean bed, and melting permafrost.

The predictions in the 2007 IPCC report are disturbing, not because of the numbers themselves, which are largely in line with forecasts in previous reports, but because the science behind them is certain enough to require action. Back in 1990, IPCC estimated that global temperatures would rise by between 0.15°C and 0.3°C per decade. Temperatures have climbed steadily since then, with the following decade and a half including the ten hottest years on record. Over this period, the rate of warming has been 0.2°C. The latest IPCC report predicts a global temperature rise over this century of anywhere between 2°C and 4.5°C, but states that 'values substantially higher than 4.5°C cannot be excluded'. Eighty per cent of the extra heat currently being trapped by the greenhouse gases is being drawn down into the oceans. However, as the oceans warm, more of that heat will remain in the air. This has the consequence that, even if emissions of greenhouse gases were to cease, the world would continue to warm for many years.

Exactly where the temperature ends up in the predicted range depends on what the greenhouse gas emissions are in coming decades. These estimates are based on a forecast that concentration of CO_2 in the atmosphere will rise from 385 to 550 ppm. Some of the uncertainty is also caused by lack of knowledge of how the global system responds to the increase in CO_2. The flow of carbon between soils, plants, the oceans, and the atmosphere is complex and still being investigated by a range of large-scale climate experiments. It is possible that, in a warmer world, ecosystems that are currently sinks for carbon, such as the Arctic tundra, may release greenhouse gases instead. Another big unknown is how the take up of carbon dioxide by the oceans might change in the future. To date, about half of our CO_2 emissions have gone into the oceans. It is difficult to predict accurately how these feedbacks between land, seas, and atmosphere will work out.

For these reasons, the global temperature increase over the coming century could be substantially higher than 4.5°C. However, even an average warming of the globe by 4°C will render the planet unrecognisable,[6] because the warming will vary around the globe. Such an intense warming will leave the Arctic free of ice, at least in the summer. Elsewhere, equally dramatic consequences are forecast. Much of the subtropics, including a region from the Mediterranean and North Africa through the Middle East to central Asia, and an area across southern Africa, are likely to experience about a 30 per cent drop in rainfall, so that the Sahara Desert could reach right into central Europe. Warmer temperatures and a severe loss of water will be felt elsewhere in the world: across China, the south western USA, Central America, most of South America, and Australia.

In Australia, there have been recent tales of thirst-crazed camels rampaging through country towns. Whether or not these stories are apocryphal, they are still indicative of the extent of the drought that the continent is already experiencing, in all probability caused by global warming. Over the past 50 years southern Australia has lost about a fifth of its annual rainfall and similar losses have been experienced in eastern Australia. Even more serious is the 70 per cent decline in the flow of Australian rivers causing a fall in the water retained by dams. This drought is also leading to reduced evaporation and transpiration as the soils get hotter. At the same time, rainfall has increased in the north-west of the continent, which seems to provide some comfort. But this respite may not last because computer models attribute this increased rainfall to the Asian haze pushing the monsoon further south. If so, as Asia cleans up its air, Australia will lose its northern rainfall.

At the same time as these droughts are occurring in the south, more northerly latitudes will get wetter as the air warms and storm tracks move. The British Isles experienced record rainfall in 2007, but it is not possible to link this single event with global warming. Storms may also strengthen and hurricanes are expected to get more intense as the oceans warm up.

Climate models assume that ice sheets melt only slowly as heat penetrates through more than 2 km of ice. However, ice sheets fracture as they melt, so that water gets rapidly to the bottom of the ice, warming its full depth and lubricating the interface between ice and bedrock. This may be the reason that the rate of ice loss in Greenland seems to have unexpectedly doubled in the past decade. The IPCC report predicts that sea level will rise by between 19 and 58 cm by 2100 because of melting ice caps and thermal expansion of the ocean. But some scientists say that, because of the effects of fracturing, this is an underestimate. Stefan Rahmstorf of Germany's Potsdam Institute for Climate Impact Research has shown that world sea levels are rising 50 per cent faster today than predicted in the last IPCC report. If the rate of sea level rise continues in line with global temperatures, he has estimated that sea levels could rise by up to 1.4 m by 2100.[7]

There have been concerns raised that the climate as a whole could have a 'tipping point' beyond which it could switch into a completely new pattern. Although there is no strong evidence for this, some individual parts of the system could be in danger of changing state quickly and perhaps irretrievably. There are three of these, all situated in the Arctic. The farthest north of these is the sea ice floating around the North Pole on the Arctic Ocean. In winter it more or less covers the Arctic Ocean basin but in summer it shrinks, creating patches of open water. The open water reflects much less sunlight than ice and so the ocean absorbs more of the summer warmth. More heat will then lead to thinner ice in the next winter, which is then easier to melt in the following summer. At present, neither observations nor models suggest that this effect will run away over the sorts of temperature ranges being predicted, but current trends in greenhouse gas emissions and global warming could eventually change this.[8]

The area of the Earth's surface north of the Arctic Circle, at only 4.5% of the globe, is too small to radically affect the planet's energy balance. A second possible tipping point is melting of the Greenland ice sheet, which has the potential to raise sea levels around the world. Melt-water making its way to the bedrock via cracks in the ice and lubricating the flow of the ice to the ocean could make the ice sheet vulnerable in a manner not included in existing climate models. Shrinkage of the ice sheet would reduce the sinking of air in the Arctic, which is an integral part of the northern Hadley cell. South of this cell are the westerlies and jet streams that drive storms around mid latitudes, and so the implications of the ice loss could extend further.[8]

Melting of sea ice and ice sheets could trigger the third tipping point: weakening of the Oceanic Global Conveyor Belt by inhibiting the formation of dense water which sinks to the ocean depths. While the climatic effects may be felt quite widely, the clearest impact could be a fall in temperatures in Europe and eastern America as the northward flow of water in the Atlantic declines. Although the Arctic does seem to be getting fresher, it does not seem to be near the danger-point. After comparing the output from 11 different ocean and climate models, Stefan Rahmstorf concluded it would take between 100,000 and 200,000 cubic metres of freshwater per second to shut down the thermohaline circulation.[9] Once it was stopped, merely reversing the conditions would not be sufficient to restart it. Adding together the effects of melting sea ice, the loss of ice from Greenland and the increased flow of rivers into the

Arctic gives barely a quarter of the lower estimate of the threshold flow, so that at present shut-down is unlikely. This could change if the melting of Greenland were to gather pace.

This then is the picture of how the planet may change by the end of this century. However, such climate forecasts and their implications are made on a macro scale, and people want to know what the effects of the weather will be where they are. The micro-scale changes, those which individuals and small populations will experience, are harder to estimate. These are the eddies accompanying the broad current flow of climate change, and they will depend on all the varied processes described in preceding chapters. How the large-scale swayings of the planet's climate are distorted as the globe warms up and how biological populations then respond will determine many of the details of how the world looks in the future. And what happens to ENSO cycles in coming decades will be especially important.

A few years ago it was thought that any future changes to the ENSO cycles were only likely to be minor.[10] However, shortly after the strong El Niño of 1997–8 results from some climate models seemed to suggest that future CO_2-induced global warming could lead to these events becoming stronger and more frequent. In addition, it has been suggested that the projected tropical warming may follow a somewhat El-Niño-like spatial pattern. This now seems less likely. After studying the predictions of several climate models, the international CMIP modelling groups found the most likely scenario is for no trend towards either El-Niño-like or La-Niña-like conditions, although there was a small probability (16%) of some tendency to El Niño conditions.[11]

There is much uncertainty in this conclusion, however. To determine what might happen to ENSO cycles in the future, William Merryfield of the Canadian Centre for Climate Modelling and Analysis looked at ENSO events in the output from 15 coupled climate models in which atmospheric CO_2 was stabilised at twice the current concentration.[12] While it seemed possible that the ENSO cycle may be minimally shortened, his conclusion was that the ENSO responses produced by the best and most up-to-date climate models were so widely scattered that they said nothing of substance about ENSO behaviour in a future warmer world. The question of how ENSO cycles will be modified by global warming therefore remains unanswered.

As the mechanisms driving the Pacific-North American (PNA) pattern and the NAO are less well understood than those involved in the ENSO phenomenon, it is hardly surprising to discover that the future of both these phenomena are equally uncertain. The effects of global warming on the PNA are especially difficult to predict because ENSO events significantly disturb the climate of the whole Pacific basin, and their future is unclear. No significant impacts of the planet's warming on the PNA have been identified, as yet. It does seem likely that the NAO will be affected by global warming, however.

During the first 100 years or so after people started collecting reliable pressure readings over the North Atlantic, the observed values of the NAO index never went up or down by too much. Then about 30–40 years ago the index started to first become more negative in the 1960s and 1970s, before swinging in the opposite direction (Fig. 7.2). This last increase corresponded to a noticeably longer growing season in Europe and milder winters in the mid Atlantic region of the United States. These stronger values of the NAO were then followed by a decline at the beginning of the new millennium. It seems that the NAO has been swaying back and forth more wildly from decade to decade during the late 20th century. By extracting a 218-year-long temperature record from Bermuda brain coral, researchers at Woods Hole Oceanographic Institution have been able to infer that this recent variability is greater than in any period as far back as the 1800s.[13] They have concluded that anthropogenic (human-related) warming does not appear to be altering whether the NAO is in a positive or negative phase, but it does seem to be increasing the NAO's variability.

Exactly how these fluctuations from decade to decade might continue to increase over the coming century is not known at present. Based on the simple model described in Chapter 7,[14] and using current estimates

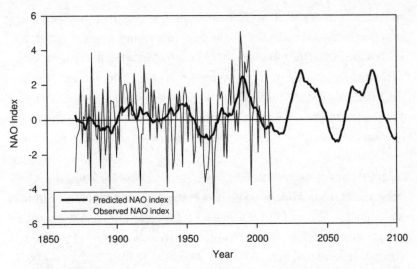

FIG 10.2 Predicted values of the NAO index for the coming century from the simple model[14] in Fig. 7.2.

of how global warming might continue, Figure 10.2 presents one possible scenario. In this picture the oscillations of recent decades continue through the century. In the absence of suitable forecasts, the predictions were made assuming the Southern Oscillation varies over the next 100 years as it did during the previous century. The model predicts a series of mild, wet winters in Europe, with cold conditions over the eastern half of the North American continent, around 2030 and 2075, with the opposite situation around the next few years and then around 2050. While the model can make definite forecasts and the results are intriguing, it still has to be borne in mind that there is no guarantee that the dynamics of the region's climate will continue to behave in the simple way they appeared to have done over the past 150 years.

Like the climate itself, the impending changes in the living world will consist of trends on a global scale combined with many much smaller geographical features, some of which may be associated with any shifts taking place in the atmospheric seesaws. In a summary of the possible impacts of future warming, the French ecologist Wilfried Thuiller[15] has pointed out that the ecological disruption wrought by climate change is generally slower than that caused by other human activities such as changes in land use, pollution, the invasion of ecosystems by alien plants and animals, and the direct effects of increased levels of carbon dioxide on plant growth. While all these human-induced changes will have consequences over the short-to-medium term, in 50 years ahead, and beyond, the effects of climate changes are likely to become even more prominent.

At the turn of the millennium, Osvaldo Sala and colleagues at the University of Buenos Aires attempted to estimate the extent of these human activities on different communities by the year 2100 from an examination of a group of terrestrial and freshwater ecosystems.[16] Overall, changes in land use were likely to constitute the main impact on populations, but the pattern was expected to vary considerably for different ecosystems. Climate change would have the strongest effect on Arctic and Alpine ecosystems, and those of the sub-Arctic and sub-Antarctic regions, whereas lakes would be more affected by invasion of non-native species. Human interventions will impact on the variety of animal and plant life present, the biodiversity. From the study, the ecosystems that were expected to experience the most significant changes to their biodiversity were those in the Mediterranean region, and also grasslands elsewhere. Northern temperate ecosystems were thought likely to suffer the least biodiversity change because major changes in land use have already happened in much of this area. Sala's group concluded by saying that changes in other communities may well depend on the interactions among the separate causes of biodiversity change. These interactions represent one of the largest uncertainties in projecting future changes in diversity. The ERSEM model simulations mentioned in Chapter 8 have already provided an illustration of interactions like this in operation by showing how an ecosystem can conspire to bring out the GSNW signal hidden in the meteorological observations.[5]

The most immediate ecological effects of climate change will be shifts in the geographical ranges of species prompted by distortions of the patterns of temperature that determine where the species can live. It has been estimated that each 1°C increase in temperature will move ecological zones northward by about 160 km. If the climate warms by 4°C over the next century, species in the northern hemisphere may have to move about 500 km closer to the Pole or else move up 500 m higher in altitude. Shifts like these have already begun to happen: the opening chapter introduced the book by describing such movements in zooplankton, fish, butterflies, birds, and whales. Higher temperatures will sometimes occur with more humid, wet conditions, but the geographical and seasonal variations of precipitation will also change. Many regions such as the Mediterranean basin will become drier in summer and this will increase stress due to droughts. The ability of species to adapt to climate change will therefore largely depend on how well they can track the shifting climate by colonising new territory, or else whether they can modify their physiology and seasonal behaviour (e.g. time of flowering or mating) to fit the new conditions in which they find themselves.

Climate change will affect different ecosystems in different ways. On land, deserts, grasslands, savannahs, and forests are each likely to differ in how they respond to changes in the annual cycles of precipitation and temperature. But some differences may also be caused by the extra carbon dioxide in the air itself. Carbon dioxide is essential for the photosynthesis of green plants, and so an increase in its atmospheric concentration will increase plant growth. Depending upon the ecosystem concerned, this may either offset or aggravate the impacts of temperature increases. The more ready availability of carbon dioxide may alter the competitive balance between species that differ in root depth or 'woodiness', as well as the underground organisms associated with them. Increased carbon dioxide has its strongest effects in regions where plant growth is limited by the availability of water.

While deserts lag behind forests and grasslands in the speed at which their plants grow, they do not do so in their response to greenhouse gases, and this is a consequence of how efficiently water is used. Rising levels of carbon dioxide appear to have led to faster tree growth in arid regions. When Xiahong Feng of Dartmouth College, New Hampshire, measured the changing ratios of different isotopes of carbon in the annual growth rings of a range of American trees, she found a trend that matched the rise in carbon dioxide levels over the past 200 years.[17] There was a steady increase in the efficiency of water use that started low in the 19th century but increased rapidly for most trees in the 20th century. This increased efficiency will be especially important in arid environments where moisture limits tree growth. It will be less so in areas whose plants are limited by the availability of nutrients or competition from other plants.

In aquatic environments, the extra carbon dioxide will have an opposite effect. When the gas dissolves in water it produces a weak acid, carbonic acid, and so increased levels of the gas in the atmosphere will raise the acidity of the oceans. This can damage the shells and skeletons of organisms, if these are made of calcium carbonate. Coral reefs with a chalky skeleton are one example; another is the 'sea butterfly' (a group of animals called Thecosomata). This animal is one of the strangest creatures in the sea, and, growing no bigger than a lentil, also one of the smallest. It is a tiny marine snail that floats and swims in large numbers in the polar seas. Its foot has developed into a pair of 'wings', hence its name. The snails are consumed by various marine species, including a wide variety of fishes that, in turn, are consumed by penguins or polar bears. Gretchen Hoffmann from the University of California, Santa Barbara, has called them the 'potato chip' of the ocean.

Comparatively little is known about the lives of sea butterflies. Sometimes they catch plankton in mucous webs, each of which, at up to 5 cm wide, is many times larger than the animal itself. At night they hunt in the surface waters before migrating down the water column in the morning. As the ocean becomes more acidic and warmer, the calcium carbonate shell protecting the animal from predators will become thinner and may disappear. It is unclear whether the creatures will be able to survive if this happens. Given their role in marine communities, the threat to the sea butterflies may have far-reaching effects.

The climate changes accompanying the global increases in carbon dioxide will affect marine ecosystems mainly through increased thermal stratification as surface waters warm up, reduced upwelling of nutrients, and the loss of sea ice. These changes will impact on phytoplankton, and the results will then percolate up the food chain from zooplankton to fish to mammals and birds. The first of these changes are starting to happen. Satellite observations have already revealed these kinds of effects on phytoplankton.

The oceans are vast and research ships are limited in number, and are sometimes limited to the speed of a bicycle. By contrast, a satellite can observe all the cloud-free areas of the globe in a few days, and can be used to estimate the concentration of phytoplankton close to the surface of the sea. This is possible because chlorophyll absorbs blue and red light more than green, chlorophyll appearing green because it reflects green light. Not that this procedure is entirely straightforward, for other processes in the atmosphere and the seawater affect the light, but with careful calibration, reliable measurements can be made. In this way, NASA's Sea-viewing Wide Field-of-view Sensor (SeaWiFS), launched in 1997, has observed phytoplankton in the ocean over more than a decade and shown how closely they follow climate changes.[18]

Links between climate and phytoplankton are clearest in the tropics and mid latitudes where there is limited vertical mixing because the ocean is thermally stratified into a warm light layer over colder, dense water. In such locations, climate warming will further inhibit mixing, thereby reducing the upward supply of nutrients from deeper down and in turn lowering productivity. Because ENSO is such a dominant mode of interannual variability and involves large-scale reorganisations of atmosphere and ocean circulations that extend around the globe, these ENSO events shine out of the observations. Phytoplankton populations were much greater during the cold phase of the El Niño of 1997–8 than during the warm period when stratification was stronger and the upwelling along the Peruvian coast was weak. However, things may be different at higher latitudes. Here, phytoplankton are often limited by low light levels because periods of intense vertical mixing can carry the cells hundreds of metres down into darkness where sunlight does not penetrate. In these regions, future warming that inhibits vertical mixing could actually increase productivity.

Globally, it seems that the abundance of phytoplankton has declined rather than increased during the past century. In 1865, Father Pietro Angelo Secchi was asked to map the clarity of the Mediterranean Sea for the Papal navy. He devised the simplest of oceanographic instruments, a 20-cm-wide white disc that is lowered until the observer loses sight of it. Determinations of Secchi depth have been a routine part of oceanographic observations ever since. They assess light penetration in the upper ocean, and can be related to phytoplankton abundance. When Daniel Boyce, Marlon Lewis, and Boris Worm of Dalhousie University in Canada combined these data with measurements of chlorophyll concentration in the upper ocean to estimate changes in phytoplankton biomass over the past 100 years, they observed a decline in eight out of ten ocean regions, a global rate of about 1% per year. While decadal fluctuations seemed to be related to basin-scale climatic indices, the long-term decline was associated with rising temperatures. As phytoplankton productivity is the base of the food web and is vital for all life in the sea, this decline does not bode well for ecosystems in a warmer world.[19]

Lakes bury the corpses of their inhabitants in the mud, and they then decay. The bacteria causing this decay consume oxygen in the process, and so during the summer months the oxygen concentration in the deepest waters of the lake, the hypolimnion, falls as dead matter rains down from above. If the lake is shallow or there is a lot of falling matter, all the deep oxygen may be used up and the deep waters instead contain gases such as hydrogen sulphide produced from the breakdown of dead cells. This does not happen in Windermere. In the future, global warming could make the bottom layers of many deep freshwater lakes, including the Great Lakes of North America, Scandinavian lakes, and some of Scotland's lochs, vulnerable to stagnation. As an example, when Ulrich von Grafenstein's group at the Laboratory of Climate and

Environmental Sciences in Gif-sur-Yvette, France, modelled how climate change would affect the circulation in two deep freshwater lakes, Lake Ammersee in Germany and Lake Annecy in France, they found that if the climate warms as predicted, the surface layers might not cool sufficiently for the layers to mix in winter and a stagnant bottom layer could develop.[20] If this happens, many deep-water animals would be in danger. Deep-dwelling crustaceans and fish could die from lack of oxygen. It is, however, difficult to say whether all this will happen: stirring up of lakes in the winter also depends on the winds, and predictions of how the winds will develop in the future are rather more uncertain. Stronger winds could break up any stagnant bottom layer.

Another expected response to the warming of the planet is for species to shift in time: plants and animals may start the year earlier or continue their seasons closer to winter. Keeping records of the times of recurring natural phenomena, such as when the first cuckoo is heard or the first blackthorn blossom is seen, is now called 'phenology', and this has a long history, predating current concerns about global warming by many centuries. In Japan and China the blossoming of cherry and peach trees is associated with ancient festivals. I have pleasant memories of taking part in the ancient ritual of picnicking among the cherry blossoms of Kyoto. Records of the dates of first blossoming extend back to the eighth century.

One of the longest European records of weather-induced events is that of Robert Marsham of Norfolk, UK, who kept a chart of different measures of the arrival of spring from 1736 onwards.[21] He meticulously catalogued detailed records of seasonal weather changes, tree foliation, crop growth, migrating birds, flowering dates of individual plants such as snowdrops and wood anemones in spring, first sightings of butterflies and swallows, and the first call of the cuckoo. He then developed these records into the 27 'Indications of Spring'. Marsham has been called Britain's first phenologist. While most were observations of the timing of natural processes, they included others such as how often—and how late in the year—the chamber pot had a skim of ice. Marsham's records did not stop with his death in 1797 but were kept up by successive observers until 1958. They then ceased, just at the point when the combined impact of burning fossil fuels and forest destruction was beginning to raise a dark cloud over the world's climate.

Inspired by Marsham's work, the UK's Royal Meteorological Society co-ordinated a nationwide network of recorders to examine the relationship between meteorological events and the natural world from 1875 to 1947. As an example, in the year 1899, there were 155 observers contributing records, and they recorded the flowering of 13 plants and the appearance of birds and insects. A summary for the year reads:

> The weather for the phenological year ending November 1899 was chiefly remarkable for its high temperatures, scanty rainfall, and splendid record of sunshine. The winter and summer were significantly warm seasons, while the autumn was also warm; but during the three spring months rather low temperatures prevailed. In the early part of the flowering season, wild plants came into blossom in advance of their mean dates, but after March they were mostly late in coming into bloom.

Reviewing all the flowering records from this scheme, Edward Jeffree reported that one significant observation was that hazel and wood anemone flowered late during the cooler springs of the 1940s. Like Hitler's army in Russia, they may have been suffering from the weak NAO in these years. A more recent British phenological study, which does cover the period when the signs of global warming began to appear, is that by Alastair and Richard Fitter at the University of York, UK.[22] They recorded the first flowering date for a set of 385 plant species in Oxfordshire over a period of 47 years. Analysis of these observations showed the average date on which these species flowered was 4.5 days earlier in the last decade of the millennium than during the previous four decades. These results are in keeping with widespread reports that trees are coming into leaf sooner, and that some typical spring flowers are increasingly being seen coming into bloom in November and December. It seems to be a biological response to warming. The authors point out that flowering is particularly sensitive to the temperature in the previous month, especially in the case of spring flowering species. The large differences in response between species shown by the study are likely to affect the structure of plant communities as climate warms: annuals are more likely to flower early than perennials, and insect-pollinated species more than wind-pollinated ones.

Phenological studies are in progress around the world, with many already seeing the signs of global warming. For more than three decades, biologist Jerram Brown from the State University of New York has trekked into the Chiricahua Mountains of southern Arizona noting on which date female Mexican jays lay their first clutch of eggs. Then, several weeks later, he has climbed up the 15-metre-tall Chihuahua pines to band each chick. His observations have revealed the jays laying their eggs earlier and earlier each season. By 1998, the first eggs of the season had arrived 10 days earlier than in 1971.[23] Brown blames global warming for speeding up the jays' reproductive clock. Although Arizona has not become particularly hotter, it has grown less cool. In the months leading up to the breeding season, average daily minimum temperatures rose by close to 3°C in the 27 years. A narrower temperature range probably encouraged earlier breeding by allowing birds to conserve energy on cold nights, when they burn off about 10% of their weight just to stay warm. The warmer air may also rouse insects earlier, which can provide extra food for females to funnel into egg production.

These kinds of changes are occurring on all continents and most oceans. A project led by Cynthia Rosenzweig of NASA's Goddard Institute for Space Studies in New York has examined 28,800 data sets looking for the impact of a changing climate on plants and animals.[24] Nine out of ten of these data sets showed alterations over the past 30 to 50 years in the direction expected as a response to global warming. Further, as in the flowering plants, the change that is having the most obvious biological impact is the early arrival of spring. Movement of the seasons, an earlier spring and later autumn, is therefore another likely consequence of the warming to come.

But all of the above are no more than the broad scale features of the biological changes to come. Accompanying these will be geographical and temporal patterns associated partly with changes in the large-scale seesaws of the climate, but also associated with the kinds of ecological responses described in Chapter 8. One particular ecological process quoted in this chapter was the vital role played by tall trees of the Canary Islands in extracting moisture from the clouds for the community below. There is already evidence that in some locations this chain is being disrupted by climate change.

Over the past 30 years, the base of the clouds that form over the northeastern states of the USA has been getting higher. Airports routinely measure the cloud ceilings for the benefit of pilots, and when Andrew Richardson and colleagues of Yale University examined data from 18 airports located along the axis of the Appalachians, they found that the cloud ceiling had risen by 180 metres or about six tree heights since 1973. The main driving force behind the change may be warmer, drier air moving up from the lowlands, which have been cleared of trees. The rise may well mean that the existing trees will not be able to scavenge water from the clouds, water on which animals such as toads and salamanders depend.[25] Elsewhere, these effects have already been seen.

Alan Pounds, an ecologist at the Monteverde Cloud Forest Preserve in Costa Rica, has documented disturbing changes on a tropical mountain in three disparate groups: amphibians, reptiles and birds.[26] Pounds and biologist Michael Fogden have connected population extinctions in 40% of the 50 species of frog, as well as range shifts in each of the three groups, to a broad climatic pattern of more mist-free days. Clouds usually hug the upper reaches of the mountains, and their mists bring moisture even without rain. Pounds has noted that the increasing frequency of dry days corresponds to rising Pacific Ocean surface temperatures. A warmer ocean heats the air, which ultimately pushes the cloud ceiling higher up the mountain. However, no one expected that the extent of the climate change that had been seen would have caused the collapse of the amphibian fauna. It seems that the higher clouds may have triggered several chains of ecological events, including one that culminated in an outbreak of chytrid fungus (a frog pathogen), that contributed to extinctions. These changes have also caused bird populations to relocate.

Clouds sailing above the tallest trees and withholding their moisture from those below is just one example of the complexities inherent in communities and the subtleties in how they may respond to changes in weather patterns, which make it difficult to foresee in any kind of detail the implications of future climate change. Consequently, the actual changes that occur at any place or time will not necessarily be those that might be intuitively expected. The experiments with the ERSEM ecosystem model have shown how the observed coupling between the plankton in the North Sea and the distant Gulf Stream could have arisen because the ecosystem was affected in a sensitive manner to weather patterns originating over the ocean, even though a clearer association with the NAO was expected. In Chapter 8 we saw that the monitoring of different ecosystems could provide lighthouses warning of impending changes. However, the species most affected, the bellwether species, will often not be known in advance, so that there will be a need to monitor a large cross-section of each community; even so, monitoring can still provide early indications of what lies ahead. In two recent investigations, plant communities, when monitored, have been shown to be just such lighthouses, warning of encroaching desert conditions.[27]

Once upon a time the Sahara desert was covered by vegetation. Then around 5,500 years ago the wet environmental conditions came to an end, plant productivity ceased and the topsoil was lost. Eventually, the green Sahara became the desert we know today. Arid ecosystems, which cover about 40% of the Earth's land area and are inhabited by over two billion people, remain vulnerable to climate change and human activities. The valleys of the rivers Tigris and Euphrates, which once nourished cities like Babylon, are becoming beset by drought and losing water due to extraction upstream. In such areas, there is always the danger of a shift from quite dry conditions to the continuous droughts of the desert. But two recent studies by Sonia Kefi from Utrecht University and colleagues working in Mediterranean ecosystems of Spain, Morocco, and Greece, and Todd Scanlon of the University of Virginia and colleagues working in the Kalahari Desert have shown that clues in the vegetation can warn of impending desertification.[27]

These groups found that, in each of the regions, the size distribution of vegetation clusters falls off as a power law. Thus, depending on the ecosystem, clusters of vegetation of a given size will be between a half and a quarter as common as clusters that are 50% smaller. Such power laws occur in other types of ecosystem and are the result of internal dynamic processes such as the interactions between nearby plants. Individual plants create local environments around themselves that minimise water run-off and facilitate the survival of other plants and seeds. Comparing satellite images from different areas in the Kalahari Basin shows that the shift towards desert conditions leads to a departure from this power law as the clusters that survive tend to be those in the less dry environments near existing trees. Increased grazing pressure has a similar impact. In the Mediterranean regions, intense cropping by animals also led to a departure from power law behaviour, with large patches becoming less and less common. Computer models showed that, in each of these study areas, these changes in size distribution can serve as warning signals of the approach of a transition to a desert state. This possibility is especially important because some of the transitions occurring in the models were both catastrophic and largely irreversible.

Observations of the large-scale distribution of plants have also been used in the forecasting of Asian monsoons. These weather events are driven mainly by contrasts between land and sea temperatures, but other factors are also at play, especially the moistness of the soils. Higher moisture content makes the transfer of heat from the land to the air through evaporation easier. Eungul Lee and his colleagues at the University of Colorado have used an analysis of satellite and other records to show that including this effect can improve monsoon forecasts.[28] Taking into account the vegetation growth—which has a strong influence on soil moisture—in the months preceding the northern and southern East

Asian monsoons allowed them to improve the reliability of forecasts by a factor of two, for the northern, and three, for the southern monsoon.

Utilising this idea that careful observations of communities may fore-shadow coming events can only be possible if the necessary monitoring programmes exist. Remote observations from satellites can provide global coverage, and projects like the continuous plankton recorder or CalCOFI programmes can generate data over geographical scales in the ocean. Many localised monitoring programmes exist as we have seen, often struggling along under the enthusiasm of a few people, but eco-logical field studies are generally of short duration, include few species, and cover small areas. Getting data spanning huge spatial and temporal scales, and involving vast numbers of species, is a tall order requiring massive reserves of people. Lucas Joppa of Duke University, North Caro-lina, has suggested another approach: citizen-science, in which qualified scientists oversee volunteers.[29]

Examples already exist. The Audubon Society's Christmas Bird Count has run for 108 years with 60,000 volunteers counting about 59 million individual birds across 1 million square kilometres of North America each year, and in Project BudBurst, participants report on the timing of botani-cal events such as flowering and leafing. Dick Schmeller at the Helmholtz Centre for Environmental Research in Leipzig and his colleagues have ana-lysed 395 citizen-science projects across five European countries, involv-ing more than 46,000 participants.[30] These volunteers donated more than 148,000 person-days per year, a figure inconceivable using professional scientists alone. Schmeller's group found the data gathered in this way are reliable and unbiased. The quality of the data was determined less by the use of volunteers, and more by survey design and the methodology employed. Joppa has argued that there is a need for a platform analogous to Wikipedia for the input, integration, mapping, and dissemination of such data.

Humanity is busily engaged in performing a hazardous global experi-ment which no scientific committee would have approved, and this is one possible approach to estimating its future implications. Forecasting the details of what might happen in the coming decades is complicated by the plethora of physical and biological processes to be considered. Mathematical models underpin much of our understanding of the world, but, as the famous fishery on the Grand Banks off Canada has shown, their predictions still cannot be totally relied upon. The models of the fisheries scientists predicted the waters would still be teeming with cod, but the fish disappeared in 1992 and show no signs of recovering.[31]

One human example of these subtle biological developments is that kidney stones will become more common. When temperatures rise, people become more dehydrated and the low urine volume leads to these calcium deposits. As a consequence, kidney stones cause much more suffering in the southern parts of the USA. Using predictions of future weather and the population trends, Tom Brikowski and his colleagues at the University of Texas have forecast that kidney stones will strike 2 million more Americans by 2050.[32]

In drawing attention to the grand patterns of change we can overlook numerous local processes. John Harte of the University of California described seeing rosy finches and ptarmigans feeding on the contracting ring of vegetation that surrounds melting snow patches on Alpine slopes, and wondered how the animals would survive if the melting occurred in the late spring rather than the early summer.[33] As he points out, his query has been answered by Robert Björk and Ulf Molau at the University of Gothenburg who have shown that release of nutrients from the snow patches sustains bryophytes, grasses, and sedges, which are likely to be replaced by shrubs and trees if the patches disappear earlier, a change which will hit herbivores hard. This is just one example of the detailed impacts of climate change that 'fall through the cracks of current, coarse projections'.

Forecasting the extent of these kinds of climatic impacts will need to be made on the basis of an understanding of how communities have behaved in the past, and so there is a continuing need to monitor ecosystems as fully as possible. Such monitoring may well show something happening at an early stage, even before the climatic change causing it becomes apparent. The connection of distant biological populations with the position of the Gulf Stream would seem to be a case in point. Here the atmospheric chain providing the linkage is not readily apparent, even though it is rooted in the pan-Atlantic seesawing of the NAO, but the biological populations seem in no doubt it is there. Similar connections might be seen happening in others of the world's oceans if the many years of observations to isolate them existed.

Another feature of the connection across the Atlantic Ocean, is that weather patterns originating over the oceans affect ecosystems some way off, often through several weather elements. Some of the plankton respond to how the oceans influence cloud cover downstream, others to the winds. This may well be an important consequence of any weakening of the thermohaline circulation and flow of the Gulf Stream resulting from global warming. The latest forecasts are for a stuttering of the sinking of water in the far north, rather than a substantial shut-down, in which case the northward flow of replacement water will be merely somewhat reduced. Detection of these changes will require careful measurements spanning the ocean, but they may also be felt by surrounding ecosystems as the weather patterns over the ocean adjust to the modified currents. Continued observation of the biological systems should reveal these changes, even if elucidating the complete mechanisms proves to be less straightforward.

Global warming will change the tempo of the intricate dance between atmosphere and oceans, and this will leave footprints in the planet's ecosystems. Only by systematically observing the world's biota can these tracks be uncovered. Such a task is vital for mankind, for these footsteps will determine how the world's living resources will look in the future.

EPILOGUE

FOOTPRINTS FROM THE DANCE

'Watch where you are putting your feet'
(Alfred Wainwright, *Memoirs of a Fellwalker*)

The riddle of the opening chapter has only been partially solved. A clearly defined ·atmospheric coupling between the numbers of plankton around the British Isles and the position of the Gulf Stream remains somewhat elusive. Yet even if the linkage was no more a fleeting similarity between a succession of weather events at the two locations, this similarity should still be visible. It is common to attribute the mild weather conditions in western Europe, such as those which allow the flora of south-west Ireland to be typical of much further south in the Continent, to the warming influence of the Gulf Stream on the air passing over it. The ocean current is cited as the reason it is possible for Inverewe Garden in northern Scotland to have many thriving, southerly plants, even though it is at the latitude of Hudson Bay and farther north than Moscow. Movements of the currents might therefore be expected to have influences around the shores of Britain. However, the mild climate is in reality down to the winds approaching the land from a relatively southerly direction and travelling over an ocean that retains heat through the winter. Any direct contribution from the currents of the Gulf Stream to this warming is not large.

The root cause of the association between the plankton and the distant ocean current is the seesaw in the region's climate known as the NAO. This is one of the great swayings that occur in the global climate, the largest of which is the ENSO phenomenon in the Pacific Ocean, whose El Niño phase wreaks havoc all around the tropics. The NAO similarly affects conditions all around the North Atlantic, and like the ENSO probably results from the ocean and the atmosphere being united in the motions of some kind of intricate dance. The atmosphere pushes the ocean, which in turn steers the way the air above it moves. As the NAO has gone through its ups and downs over the last few decades, it has been driving both the path of the Gulf Stream and the lives of the distant plankton. But the timescales do not match: the ocean current lumbers along slowly, taking years to adjust its position, while the zooplankton live and die in a few weeks, having little memory of the previous year's weather. How do the zooplankton go up and down with the slow pace of the ocean current? Could it be that the ocean circulation is, in some way, the source of the weather patterns to which the plankton populations respond. Then the weather experienced by the animals would change from year to year at a rate similar to that of the ocean currents.

Unfortunately, in keeping with the minor role of the ocean currents in the amelioration of Europe's weather, all the evidence from both the North Atlantic and North Pacific Oceans is that any such effect must be weak. The ocean's influence will not stand out markedly in the weather variability that comes in from around the world. While the ocean marches at the atmosphere's bidding, its footsteps leave only weak imprints on the atmosphere in return. Even so, the oceanic effects accompanying the shifts in the Gulf Stream still appear in the eastern ecosystems. This implies that living communities can often pick up such effects that are embedded in the weather patterns. Both observations and modelling studies seem to bear this out. Maize crops in Zimbabwe appear to be even more sensitive to ENSO events than the rainfall on which they depend because the arable system and its management were influenced by more than this single meteorological variable. In the ERSEM model, the planktonic ecosystem tracked the Gulf Stream because an ecosystem responded as a whole to good or bad weather.

This sensitive response of ecosystems to comparatively weak but systematic changes in the weather, especially those originating over the oceans, may be a widespread feature. There are several ways through which ecosystems can show significant shifts in response to moderate changes in the climate: they may be very dependent on one species, a mismatch of timing may disengage an important animal population from its food, one species may be an ecological bellwether presaging future trends, or the interaction of different members of the community may bring out a common signal from the welter of weather fluctuations, as a distant aerial view reveals a buried structure in the wealth of vegetation seen on the ground. Monitoring as many communities as possible therefore has the potential of giving warnings of the climate changes that are ahead.

There is no longer any doubt that global warming is coming. We are continuing to release greenhouse gases into the atmosphere and the general impact of these gases is straightforward to predict. These effects are already being seen. However, while the broad sweep of the future can be predicted, the detailed and local changes are less easy to foresee. These local effects will depend on how the global seesaws in the climate like the ENSO and the NAO fluctuate in the warmer world, and this remains unknown. If some ecosystems respond to the initial stages of shifting weather patterns, they could provide an early indication of how some part of the climate is developing. Only by continuous careful monitoring can these lighthouses reveal such changes and warn us of trends, but some of these potential warning lights are only surviving on a hand-to-mouth basis. But once shifts in climatic patterns are noticed, it still remains to be seen how far our course can then be corrected. One possible approach to ensuring future conditions that might be more navigable even touches on the coupling with the Gulf Stream.

Among the biological populations which have been observed tracking the Gulf Stream, are the wild flowers studied by Arthur Willis on the roadside verge near Bibury, which were described in Chapter 3. Some of the analysis of these data was carried out by Nigel Dunnett of Sheffield University, UK. Nigel was also involved in making some of the original observations and, since Arthur's death, has taken on the task of making the annual visits to Akeman Street. But Nigel Dunnett's research has another dimension, inspired in part by the impact of the warming climate at Bibury: in 2009, he won a silver-gilt medal for a garden that he exhibited at the world-renowned Chelsea Flower Show.

His *Future Nature Garden* was designed to be a new type of drought-resistant urban garden. By careful plant selection, using some simple planting methods and avoiding irrigation except by stored rainwater, it aimed to both alleviate the pressure on urban drainage infrastructure in wet weather and to optimise the use of water during increasingly dry summer months. One key feature of the garden was a green roof to help reduce surface water runoff, while at the same time enhancing biodiversity. The garden was intended to demonstrate principles that can be applied to private gardens, community spaces, schools, factories, office developments, and urban housing estates.

Nigel's ideas are merely one small approach to ameliorating what lies ahead. There are a multitude of other such developments around the globe, most of which are much larger in scale and scope. They range from more efficient ways of using energy to alternative sources of energy. These include the use of solar heating, winds, waves and tides, nuclear power, and biofuels. And then there are a range of methods for capturing and disposing of greenhouse gases. There are problems to be overcome with each of these proposals but together they attempt to point a positive path into the coming century, providing we take care with our footsteps.

ENDNOTES

Chapter 1

1.   Egede, H., 1745. *History of Greenland* (translated from the Danish), Pickering Book-seller, Piccadilly, London. Quoted in Hurrell, J.W. *et al.*, 2003. The North Atlantic Oscillation Climatic Significance and Environmental Impact, Geophysical Monograph **134**, 37–50.

2.   Taylor, A.H., 1995. North–south shifts of the Gulf Stream and their climatic connection with the abundance of zooplankton in the UK and its surrounding seas, *ICES J. Mar. Sci.*, **52**, 711–721; Taylor, A.H., Colebrook, J.M., Stephens, J.A., and Baker, N.G., 1992. Latitudinal displacements of the Gulf Stream and the abundance of plankton in the north-east Atlantic. *J. Mar. Biol Ass.*, **72**, 919–921. The Gulf Stream north wall (GSNW) data are available at: www.pml.ac.uk/gulfstream.

3.   These data are from a letter by Benjamin Franklin, drafted at sea, aboard the London packet *Capt.Truxton*, and published in the *Philosophical Transactions of the Royal Society* in, 1785. The letter contains a range of ruminations on the science of sailing ships and sea crossings.

4.   The methods are described in Taylor, A.H. and Stephens, J.A., 1980. Latitudinal displacements of the Gulf Stream (1966 to 1977) and their relation to changes in temperature and zooplankton abundance in the NE Atlantic, *Oceanol. Acta*, **3(2)**, 145–149; and also Taylor 1995 (ibid). The north wall of the Gulf Stream has been studied in several later studies: Miller, J.L., 1994. Fluctuations of Gulf Stream position between Cape Hatteras and the Straits of Florida. *J. Geophys. Res.*, **99(C3)**, 5057–5064; Drinkwater, K.F., Myers, R.A., Pettipas, R.G., and Wright, T.L., 1994. Climatic data for the north-west Atlantic: the position of the shelf/slope front and the northern boundary of the Gulf Stream between 50°W and 75°W, 1973–1992. *Can. Data Rep. Hydrogr. Ocean Sci.*, **125**, 103p; Joyce, T.M., Deser, C., and Spall, M.A., 2000. On the relation between decadal variability of subtropical Mode Water and the North Atlantic Oscillation, *J. Climate*, **13**, 2550–2569.

5.   This was first reported in Curry R.G. and McCartney, M.S., 2001, Ocean gyre circulation changes associated with the North Atlantic Oscillation. *Journal of Physical Oceanography*, **31**, 3374–3400.

6.   This summary is taken from the detailed descriptions in: Glover, R.S., 1967. The continuous plankton recorder survey of the North Atlantic. *Symposia of the Zoological*

Society of London, **19**, 189–210; and Warner, A.J. and Hays, G.C., 1996. Sampling by the continuous plankton recorder survey. *Progress in Oceanography*, **34**, 237–256.

7.   Hardy, A.C., 1967. *Great Waters*, Collins, London, pp542. This book contains many of his fine watercolour paintings of the Southern Ocean.

8.   These developments have been described in his book: Hardy, A.C., 1971 (first published in 1956). *The Open Sea, The World of Plankton*, Collins.

9.   Dickson, R.R., 1992. The natural history of time-series. In *Dynamical Systems Theory for Ecosystems*, pp. 70–98. Bob Dickson has drawn an important conclusion from this analysis: 'But it is nevertheless worth emphasising to funding agencies that adding another year to a 60-year record may be inherently more valuable than starting anew elsewhere.'

10.   Taylor *et al.*, 1992 (ibid); Hays, G.C., Carr, M.C., and Taylor, A.H., 1993. The relationship between Gulf Stream position and copepod abundance derived from the continuous plankton recorder survey: separating biological signal from sampling noise. *J. Plankt. Res.*, **15**, 1359–1373.

11.   Leibniz, G.W. (Translated by C. Cohen and A. Wakefield), 2008. *Protogaea*, University of Chicago Press, pp. 204.

12.   Taylor 1995 (ibid); Hays *et al.* 1993 (ibid).

13.   These other linkages have been reported by: Frid, C.L.J. and Huliselan, N.V., 1996. Far field control of long-term changes in Northumberland (NW North Sea) coastal zooplankton. *ICES J. Mar. Sci.*, **53(6)**, 972–977; George, D.G. and Taylor, A.H., 1995. UK lake plankton and the Gulf Stream, *Nature*, **378**, 139; George, D.G., 2000. The impact of regional-scale changes in the weather on long-term dynamics of *Eudaptomus* and *Daphnia* in Esthwaite Water, Cumbria. *Freshwater Biology*, **45**, 111–121; Jennings E. and Allott, N., 2006. Position of the Gulf Stream influences lake nitrate concentrations in SW Ireland. *Aquatic Sci.*, **68**, 482–489; Willis, A.J., Dunnett, N.P., Hunt, R., and Grime, J.P., 1995. Does Gulf Stream position affect vegetation dynamics in western Europe? *Oikos*, **73**, 408–410; Ottersen, G., Ådlandsvik, B., and Loeng, H., 2000. Predicting the temperature of the Barents Sea. *Fisheries Oceanography*, **9(2)**, 121–135.

14.   Segarin, R. and Micheli, F., 2001. Climate change in non-traditional data sets. *Science*, **294**, 811.

15.   Parmesan, C. *et al.*, 1999. Poleward shifts in geographical ranges of butterfly species associated with regional warming. *Nature*, **399**, 579–583.

16.   Beaugrand, G., Reid, P.C., Ibanez, F., Lindley, J.A., and Edwards, M.E., 2002. Reorganization of North Atlantic marine biodiversity and climate. *Science*, **296**, 1692–1694.

17.   Stebbing, A.R.D., Turk, S.M.T., Wheeler, A., and Clarke, K.R., 2002. Immigration of southern fish species to south-west England linked to warming of the North Atlantic (1960–2001). *J. Mar. Biol. Ass. UK*, **82**, 177–180.

18. Grebmeier, J.M., Overland, J.E., Moore, S.E., Farley, E.V., Carmack, E.C., Cooper, L.W., Frey, K.E., Helle, J.H., McLaughlin, F.A., and McNutt, S.L., 2006. A major ecosystem shift in the northern Bering Sea. *Science*, **311**, 1461.

19. Warren, M.S., Hill, J.K., Thomas, J.A., Asher, J., Fox, R., Huntley, B., Roy, D.B., Telfer, M.G., Jeffcoate, S., Harding, P., Jeffcoate, G., Willis, S.G., Greatorex-Davies, J.N., Moss, D., and Thomas, C.D., 2001. Rapid responses of British butterflies to opposing forces of climate and habitat change. *Nature*, **414**, 65–68.

20. Reid, P.C., Edwards, M.E., Hunt, H.G., and Warner, A.J., 1998. Phytoplankton change in the North Atlantic. *Nature*, **391**, 546.

21. Myneni, R.B., Keeling, C.D., Tucker, C.L., Asrar, G., and Nemar, R.R., 1997. Increased plant growth in the northern latitudes from 1981 to 1991. *Nature*, **386**, 698–702.

22. Aebischer, N.J., Coulson, J.C., and Colebrook, J.M., 1990. Parallel long-term trends across four marine trophic levels and weather. *Nature*, **347**, 753–755.

23. Duarte, C.M., Cebrian, J., and Marba, N., 1992. Uncertainty of detecting sea change. *Nature*, **356**, 190. In this paper, the authors show that, although there has been an exponential increase in the initiation of new monitoring programmes generating time-series of observations by European marine stations during recent decades, there has also been a similar increase in their termination. The net result is that 'long-term monitoring programmes are, paradoxically, among the shortest projects in marine sciences: many are initiated but few survive a decade.'

24. Pierce, F., 2008. Ozone hole? What ozone hole? *New Scientist*, **199**, Sept. 20, 46–47; Farman, J.C., Gardiner, B.G., and Shanklin, J.D., 1985. Large losses of total ozone in Antartica reveal seasonal ClO_x/NO_x interaction. *Nature*, **315**, 207–210.

Chapter 2

1. Huntford, R., 1997. *Nansen: The Explorer as Hero*. Duckworth, 610pp.

2. Gaskell, T.F., 1972. *The Gulf Stream*. Cassell and Co. Ltd, 170pp.

3. Richardson, P.L., 1980. Benjamin Franklin and Timothy Folger's first printed chart of the Gulf Stream. *Science*, **207**, 643–645.

4. Richardson, P.L., 1985. Drifting derelicts in the North Atlantic 1883–1902. *Progress in Oceanography*, **14**, 463–483.

5. Ebbesmeyer, C and Ingraham, W.J., 1994. Pacific toy spill fuels ocean current pathways research. *Earth in Space*, **7(2)**, 7–9, 14; *The Times*, 28 June, 2007; Ebbesmeyer, C and Scigliano, E., 2009. *Flotsametrics and the Floating World: How One Man's Obsession with Runaway Sneakers and Rubber Ducks Revolutionised Ocean Science*. Harper Collins, 304pp.

6.   Moore, C.J., Moore, S.L., Leecaster, M.K., and Weisberg, S.B., 2001. A comparison of plastic and plankton in the North Pacific Central Gyre. *Marine Pollution Bulletin*, **42 (12)**, 1297–1300.

7.   Kunzig, R., 2000. *Mapping the Deep: The Extraordinary Story of Ocean Science*. Sort Of Books, London, 345pp.

8.   Stommel's work and the history of the Gulf stream are described in his book: Stommel, H., 1958. *The Gulf Stream*. University of California Press, 202pp.

9.   The Parsons–Veronis model is described in: Parsons, A.T., 1969. A two-layer model of Gulf Stream separation. *J. Fluid. Mech.* **39**, 511–528 and Veronis, G., 1973. Model of world ocean: I. Wind-driven, two-layer. *J. Mar. Res.* **31**, 228–288; and the observational test in: Gangopadhyay, A., Cornillon, P., and Watts, R.D., 1992. A test of the Parsons–Veronis hypothesis on the separation of the Gulf Stream. *J. Phys. Oceanogr.*, **22**, 1286–1301, and Cornillon, P. and Gangopadhyay, A., 1991. Why does the Gulf Stream leave the coast at Cape Hatteras? *Maritimes*, February, 13–15.

10.   Taylor, A.H. and Gangopadhyay, A., 2001. A simple model of interannual displacements of the Gulf Stream. *Journal of Geophysical Research*, **106(C7)**, 13849–13860. This model is based on that in: Behringer, D., Regier, L., and Stommel, H., 1979. Thermal feedback on wind-stress as a contributing cause of the Gulf Stream. *Journal of Marine Research*, **37**, 699–709. In this model, the ocean is represented by a single string of boxes running from north to south.

11.   Gulf Stream meandering and ring formation are described in: Cornillon, P., 1992. Gulf Stream. *Encyclopedia of Earth Science*, **2**, 465–480; Watts, D.R., 1983. Gulf Stream variability. In *Eddies in Marine Science*, (ed. A.R. Robinson), Springer-Verlag, Berlin, Heidelberg, 114–144 and Watts, D.R., 1991. A synoptic view of the Gulf Stream. *Maritimes*, February, 3–6.

12.   Richardson, P.L., 1993. A census of eddies observed in North Atlantic SOFAR float data. *Progress in Oceanography*, **31**, 1–50.

13.   Stommel, H., 1987. *A View of the Sea*. Princeton University Press, Princeton. New Jersey, 165pp.

14.   The conveyor belt is described in: Bigg, G.R., 1996. *The Oceans and Climate*. Cambridge University Press, 266pp; and also in Kunzig 2000 (see note 7 above).

15.   Burroughs, W.J., 2001. *Climate Change: A Multidisciplinary Approach*. Cambridge University Press, 298pp.

16.   Manabe, S. and Stouffer, R.J., 1993. Century-scale effects of increased CO_2 on the ocean-atmosphere system. *Nature*, **364**, 215–218.

17.   Toggweiler, J.R. and Russell, J., 2008. Ocean circulation in a warming climate. *Nature*, **451**, 286–288.

18.   Donnelly, J.P., Driscoll, N.W., Uchupi, E., Keigwin, L.D., Schwab, W.C., Thieler, E.R., and Swift, S.A., 2005. Catastrophic meltwater discharge down the Hudson Valley: a potential trigger for the Intra-Allerød cold period. *Geology*, **33**, 89–92.

19.   Simons, P., 1997. *Weird Weather*. Warner Books, 308pp.

20.   Svensmark, H. and Calder, N., 2007. *The Chilling Stars, A Cosmic View of Climate Change*. Icon Books, pp 268.

21.   Rennell, J., 1832. *An investigation of the currents of the Atlantic Ocean and of those which prevail between the Indian and the Atlantic Ocean*. (ed. J. Purdy), J.G. and F. Rivington, London.

22.   Lamb, H.H., 1959. Our changing climate, past and present. *Weather*, October, 299–318.

Chapter 3

1.   Carl Linnaeus' story is taken from: Knapp, S., 2000. What's in a name? Linnaeus' marginal jottings created order out of botanical chaos. *Nature*, **408**, 33.

2.   Jones, S., 2000. *Almost Like a Whale*. Anchor, Transworld Publishers, 499pp. There are many other examples. *The Wildlife Companion* (ed. M. Tait and O. Taylor), Robson Books, 2004 lists the following: *Agra cadabra* (a carabid), *Apopyllus now* (a spider), *Cyclocephala nodanotherwon* (a scarab beetle), *Ittibittium* (a tiny mollusc), *Heerz lukenatcha* (a braconid), *Kamera lens* (a protist), *La cucaracha* (a pyralid), *Pieza kake* (a fly), *Verae peculya* (a horse fly), *Vini vidivici* (a parrot), and *Ytu brutus* (a water beetle).

3.   The description of the Competitive Exclusion Principle is based on: Colinvaux, P., 1978. *Why Big Fierce Animals Are Rare*. George Allen and Unwin Ltd, London, Boston, Sydney, 224pp.

4.   George, D.G. and Harris, G.P., 1985. The effect of climate on long-term changes in the crustacean zooplankton biomass of Lake Windermere, UK. *Nature*, **316**, 536–539.

5.   George, D.G. and Taylor, A.H. 1995. UK lake plankton and the Gulf Stream, *Nature*, **378**, 139.

6.   George, D.G., 2000. The impact of regional-scale changes in the weather on long-term dynamics of *Eudaptomus* and *Daphnia* in Esthwaite Water, Cumbria. *Freshwater Biology*, **45**, 111–121.

7.   Much of the history of the FBA and the perch programme is described in Fryer, G., 1991. *A Natural History of the Lakes, Tarns and Streams of the English Lake District*. Freshwater Biological Association, 368pp. and Pickering, A.D., 2001. *Windermere: Restoring the Health of England's Largest Lake*. Freshwater Biological Association Special Publication No. 11, 126pp. These books also describe the ecology of Windermere and other lakes.

8. Haugen, T.O. *et al.*, 2006. The ideal free pike: 50 years of fitness maximizing dispersal in Windermere. *Proc. Roy. Soc. Lond.* doi:10.1098/rspb.2006.3659; Conover, D.O., 2007. Nets versus nature. *Nature*, **450**, 179–180; Edeline, E. *et al.* Trait changes in a harvested population are driven by a dynamic tug-of-war between natural and harvest selection. 2007. *Proceedings of the National Academy of Science USA*, **104**, 15799–15804.

9. Arthur Hassall's life has been described in: Laidlaw, E.F., 1990. *The Story of the Royal National Hospital, Ventnor*, Crossprint, Newport, Isle of Wight, 133pp and in Philipsborn, C., 2006. Pioneer still leaves us with food for thought. *Isle of Wight County Press*, December 22, p. 3.

10. William Smith's story is told in: Winchester, S., 2002. *The Map that Changed the World*, Penguin Books, 338pp.

11. Lewis, D.H., 2006. Obituary of Professor Arthur Willis. *The Independent*, 15 July.

12. Willis, A.J., Dunnett, N.P., Hunt, R., and Grime, J.P., 1995. Does Gulf Stream position affect vegetation dynamics in Western Europe? *Oikos*, **73**, 408–410.

13. Dunnett, N.P., Willis, A.J., Hunt, R., and Grime, J.P., 1998. A 38-year study of relations between weather and vegetation dynamics in road verges near Bibury, Gloucestershire. *Journal of Ecology*, **86**, 610–621.

14. Dunnett, N.P. and Grime, J.P., 1999. Competition as an amplifier of short-term vegetation responses to climate: an experimental test. *Functional Ecology*, **13**, 388–395.

15. Niering, W.A., Wuittaker, R.H. and Lowe, C.H., 1963. The Saguaro: a population in relation to its environment. *Science*, **142**, 15–23.

16. Jones, S., 2000. *Almost Like a Whale*, Transworld Publishers, 499pp.

17. Thornton, I., 1996. *Krakatau: The Destruction and Reassembly of an Island Ecosystem*. Harvard University Press, 327pp.

Chapter 4

1. MacLulich, D.A., 1937. Fluctuations in the numbers of the varying hare (*lepus americanus*). University of Toronto Studies Biological Series 43. University of Toronto Press, Toronto; Elton, C. and Nicholson, M., 1942. The ten-year cycle of the lynx in Canada. *J. Animal Ecology*, **11**, 215–244.

2. Odum., 1957. *Fundamentals of Ecology*. Saunders.

3. John Scott Russell's story is described in: Darrigol, O., 2003. The Spirited Horse, the Engineer, and the Mathematician: Water Waves in 19th-Century Hydrodynamics. *Arch. Hist. Exact Sci.* **58**, 21–95; and in Craik, A.D.D., 2004. 'The origins of water wave theory'. *Annual Review of Fluid Mechanics*, Vol. **36**, pp. 1–28.

4. Russell, John Scott, 1845. Report on Waves: Report of the fourteenth meeting of the British Association for the Advancement of Science, York, September 1844. (London, pp. 311–390, Plates XLVII–LVII).

5. Tamura, H., Waseda, T., and Miyazama, Y. Freakish sea state and swell wind-sea coupling: numerical study of the *Suwa-Maru* incident., 2009. *Geophys. Res. Letts.* Doi:10.129/2008GL036280; Anon., 2009. Rogue waves. *Nature*, **457**, 514.

6. Holliday, N.P., Yelland, M.J., Pascal, R.W., Swail, V.R., Taylor, P.K., Griffiths, C.R., and Kent, E.C., 2006. Were extreme waves in the Rockall Trough the largest ever recorded. *Geophysical Research Letters*, **33**, L05613.

7. Deterministic chaos and the limits it imposes on predictability are described more fully in: Gleick, J., 1989. *Chaos: the Making of a New Science.* Viking, 353pp.

8. Thompson, D.W., 1992. *On Growth and Form.* Dover reprint of 1942 2nd edition (1st edition 1917).

9. Guth, A.H., 1998. *The Inflationary Universe.* Vintage, 358pp.

10. Bell, E.T., 1965. *Men of Mathematics* 2. Penguin Books, 646pp.

11. Quoted in: Gell-Mann, M., 1994. *The Quark and the Jaguar: Adventures in the Simple and the Complex.* Abacus, 392pp.

12. Laskar, J. and Gastineau, M., 2009. Existence of collisional trajectories of Mercury, Mars and Venus with the Earth. *Nature*, **459**, 817–819.

13. May, R.M., 1976. Simple mathematical models with very complicated dynamics. *Nature*, **261**, 459–467.

14. Stenseth, N.C. and Ims, R.A. (eds)., 1993. *The Biology of Lemmings.* Academic Press, London; Coulson, T. and Malo, A., 2008. Case of the Absent Lemmings. *Nature*, **456**, 43–44.

15. Colinvaux, P., 1978. *Why Big Fierce Animals Are Rare.* George Allen and Unwin Ltd., London, Boston, Sydney, 224pp.

16. May, R.M., 1973. *Stability and Complexity in Model Ecosystems.* Princeton University Press.

17. Hassall, M.P., Lawton, J.H., and May, R.M., 1976. *Journal of Animal Ecology*, **46**, 471–486.

18. Tilman, D., Reich, P.B., and Knops, J.M.H., 2006. Biodiversity and ecosystem stability in a decade-long grassland experiment. *Nature*, **441**, 629–632.

19. Beninca, E, Huisman, J., Heerkloss, R., Johnk, K.D., Branco, P., Van Nes, E.H., Scheffer, M., and Ellner, S.P., 2008. Chaos in a long-term experiment with a plankton community. *Nature*, **451**, 822–825.

20. Berryman, A.A. and Millstein, J.A., 1989. Are ecological systems chaotic – and if not, why not? *Trends Ecol. Evol.*, **4**, 26–29.

21. Taylor, A.H. and Joint, I.R., 1990. A steady-state analysis of the 'microbial loop' in stratified systems. *Mar. Ecol. Prog. Ser.*, **59**, 1–17.

22. Neutal, A-M, Heesterbeek, J.A.P., van de Koppel, J., Hoenderboom, G., Vos, A., Kalderway, C., Berendse, F. and de Ruiter, P.C., 2007. Reconciling complexity with stability in naturally assembling food webs. *Nature*, **449**, 599–602.

23. Lovelock, J., 2000. *Gaia: A New Look at Life on Earth*. Oxford University Press.

24. Charlson, R., Lovelock, J., Andreae, M., and Warren, S., 1987. Oceanic phytoplankton, atmospheric sulphur, cloud albedo and climate. *Nature*, **326**, 655–661.

25. Jones, G., 2005. *Marine and Freshwater Res.*, **55**, 849; George, A., 2005. Coral reefs create clouds to control climate. *New Scientist*, 5 February.

26. Lefèvre, N, Taylor, A.H., and Geider, R.J., 2001. Phytoplankton physiology can affect ocean surface temperatures. *Geophysical Research Letters*, **28**, 1251–1254.

27. Kausrud, K.L., Mysterud, A., Steen, H., Vic, J.O., Ostbye, E., Cazelles, B., Framstad, E., Eikeset, I., Mysterud, I., Solhay, T., and Stenseth, N.C., 2008. Linking climate change to lemming cycles. *Nature*, **456**, 93–97.

Chapter 5

1. Costantino, M., 2005. *Weather Handbook: A Guide to the Earth's Weather and Climate*. Silverdale Books.

2. Halford, P., 2005. *Storm Warning: The Origins of the Weather Forecast*. Sutton Publishing, 295pp.

3. Hardy, A.C., 1967. *Great Waters*, Collins, London, 542pp.

4. Stamp, T. and Stamp, C., 1982. William Scoresby Junior (1789–1857). *Arctic*, v. 35, no. 4, Dec. 1982, pp. 550–551, 1 port. ASTIS record 32555.

5. Hamblyn, R., 2001. *The Invention of Clouds: How an Amateur Meteorologist Forged the Language of the Skies*. Picador, 291pp.

6. Hearn, Chester G., 2002. *Tracks in the sea: Matthew Fontaine Maury and the mapping of the oceans*. Camden, Maine: International Marine. ISBN 0071368264.

7. FitzRoy, R.A., 1860. On British Storms. *Report of the British Association for the Advancement of Science*, 39–44; FitzRoy, R.A., 1863. *The Weather Book: A Manual of Practical Meteorology*, Longman, Green, Longman, Roberts and Green, London.

8. FitzRoy's life and achievements have been described by: Halford (see note 2 above).

9. Atkinson, B.W. and Gadd., A., 1986. *A Modern Guide to Weather Forecasting*, Mitchell Beazley, London, 160pp.; Smith, A., 2000. *The Weather*, Arrow Books, 285pp.

10. Egede, H., 1745. *History of Greenland* (translated from the Danish), Pickering Bookseller, Picadilly, London. Quoted in Hurrell, J.W. *et al.*, 2003. The North Atlantic Oscillation Climatic Significance and Environmental Impact, *Geophysical Monograph* **134**, 37–50. Hans Egede's son Poul reported one of the most famous sea-serpent sightings. According to his report, on the 6 July 1734 he saw a 'most dreadful monster' off the coast of Greenland. The head of this enormous creature reached up to the yardarm (Greener, M. 2010 The golden age of sea-serpents, *Fortean Times*, April, 32–38).

11. Hurrell, J.W., 1995. Decadal trends in the North Atlantic Oscillation: regional temperatures and precipitation. *Science*, **269**, 676–679.

12. van Loon, H. and Rogers, J.C., 1978: The Seesaw in Winter Temperatures between Greenland and Northern Europe. Part I: General Description. *Monthly Weather Review*, **106**, 296–310.

13. Taylor, A.H. and Stephens, J.A., 1998. The North Atlantic Oscillation and the latitude of the Gulf Stream. *Tellus*, **50A**, 134–142.

14. Taylor, A.H. and Gangopadhyay, A., 2001. A simple model of interannual displacements of the Gulf Stream. *Journal of Geophysical Research*, **106(C7)**, 13849–13860. This model is based on that in: Behringer, D., Regier, L., and Stommel, H., 1979. Thermal feedback on wind-stress as a contributing cause of the Gulf Stream. *Journal of Marine Research*, **37**, 699–709.

15. There are no other observations that can be used to test the simple box model. However, it can be compared with the predictions from one of the most complex models of how the atmosphere and ocean interact in the climate system. Figure E5.1 shows that the BRS model is in agreement with 100 years of results from one such model, the UK Hadley Centre's HadCM3. This model reproduces many aspects of the effects of the NAO on the Gulf Stream (de Coëtlogen, G., Frankignoul, C., Bentsen, M., Delon, C., Haak, H., Masina, S., and Pardaens, A. 2006. Gulf Stream variability in five oceanic general circulation models. *J. Phys. Oceanogr.*, **36**, 2119–2135).

16. Hameed, S. and Piontkovski, S., 2004. The dominant influence of the Icelandic low on the position of the Gulf Stream northwall. *Geophysical Research Letters*, **31**, L09303, doi:10.1029/2004GL015561.

Chapter 6

1. Gandhi, M.K. *The Collected Works of Mahatma Gandhi Vol II*, Wikisource; Bohr, P.R., 1972. *Famine in China and the missionary: Timothy Richard as relief administrator and advocate of national reform, 1876–1884*.

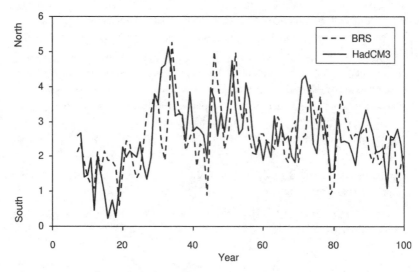

FIG E5.1 Latitude at which the Gulf Stream leaves the US coast as predicted by the Behringer, Regier and Stommel model compared with 100 years of predictions from the climate model HadCM3 (the graphs are in standardised units).

2.   Walker, G.T., 1923. Correlation in seasonal variation of weather, VIII, a preliminary study of world weather. *Mem. Ind. Met. Dept.*, **24**, 73–131.

3.   Couper-Johnston, R., 2000. *El Niño The Weather Phenomenon That Changed the World*, Hodder and Stoughton, 354pp.

4.   Suplee, C., 1999. El Niño/La Niña: nature's vicious cycle. *National Geographic*, **195**, 72–95.

5.   Fagan, B., 2004. *The Long Summer. How Climate Changed Civilization*. Grants Books, London, 284pp.

6.   Ballabrera-Poy, J.K., Murtugudda, J.R., and Busalacchi, A.J., 2002. On the potential impact of sea surface salinity observations on ENSO predictions. *J. Geophys. Res.*, **107**, 8007; Hecht, J., 2002. Sinking salt heralds next El Niño. *New Scientist.*

7.   Thompson, J., 1901. Nineteenth-century clouds over the dynamical theory of heat and light. *The London, Edinburgh and Dublin Philosophical Magazine and Journal of Science*, Series **6, 2,** 1.

8.   Anon., 2005. El Niño helped beetles beat invader. *New Scientist*, 4 June, 19; Williams, A., 2005. *Aquatic Botany*, DOI:10.1016/j.aquabot.2005.01.003.

9.   Kumar, K.K., Rajagopalan, B., Hoerling, M., Bates, G., and Cane, M., 2006. Unraveling the mystery of Indian Monsoon failure during El Niño. *Science*, doi:10.1126/science1131152.

10.   Fromentin, J-M. and Planque, B., 1996. *Calanus* and environment in the eastern North Atlantic. 2. Influence of the North Atlantic Oscillation on *C. finmarchicus* and *C. helgolandicus*. *Marine Ecology Progress Series*, **134**, 111–118.

11.   Stephens, J.A., Jordan, M.B., Taylor, A.H., and Proctor, R., 1998. The effects of fluctuations in North Sea flows on zooplankton abundance. *Journal of Plankton Research*, **20**, 943–956.

12.   Attrill, M.J. and Power, M., 2002. Climatic influence on a marine fish assembly. *Nature*, **417**, 275–278.

13.   Elliott, J.M., 1975. The growth of brown trout (*Salmo trutta* L.) fed on reduced rations. *Journal of Animal Ecology*, **44**, 823–842; Taylor, A.H., 1978. An analysis of the trout fishing at Eye Brook – a eutrophic reservoir. *Journal of Animal Ecology*, **47**, 407–423.

14.   Taylor, A.H., Jordan, M.B., and Stephens, J.A, 1998. Gulf Stream shifts following ENSO events. *Nature*, **393**, 638.

15.   Pearce, F., 2009. North Atlantic is the big cheese of global climate. *New Scientist*, 14 Feb., 14; Tsonis, A., 2009. *Geophys. Res. Letts.*, DOI:10.1029/2008GL036874.

Chapter 7

1.   Sabine, E., 1846. On the causes of remarkably mild winters which occasionally occur in England. *Philosophical Magazine and Journal*, 317–324.

2.   Gerrit de Veer., 1876. 'The Three Voyages of William Barents to the Arctic Regions'.

3.   Anon., 1981. New Light on Novaya Zemlya Polar Mirage. *Physics Today*, 34: 21, January.

4.   The method of predicting temperatures was developed by: Ottersen, G., Ådlandsvik, B, and Loeng, H., 2000. Predicting the temperature of the Barents Sea. *Fisheries Oceanography*, **9(2)**, 121–135; and the drifting of warm- and cold-water patches was examined in: Taylor, A.H., 1983. Fluctuations in the surface temperature and surface salinity of the north-east Atlantic at frequencies of one cycle per year and below. *J. Climatol.*, **3**, 253–269.

5.   Morgan, P.J., 1990. *The Leatherback Turtle: Sea Turtles and their Conservation*. National Museum of Wales, Cardiff, 34pp.

6.   Minobe, S., Kuwano-Yoshida, A., Komori, N., Xie, S.-P., and Small, R.J., 2008. Influence of the Gulf Stream on the Troposphere. *Nature*, **452**, 206–210.

7.   Fontaine Maury, M., 1856. *The Physical Geography of the Sea and its Meteorology*; FitzRoy, R., 1959. *Notes on Meteorology*. Board of Trade Miscellaneous Paper, HMSO, London.

8. Stommel, H., 1958. *The Gulf Stream*. University of California Press, 202pp.

9. Connor, S., 2003. Forget about the Gulf Stream myth: Britain is kept warm in winter by the Rocky Mountains. *The Independent*, 10 February; Seager, R., Battisti, D.S., Yin, J., Gordon, N., Naik, N., Clement, A.C., and Cane, M.A., 2002. *Quart. J. Roy. Met. Soc.*, **128**, 2563–2586.

10. Sawyer, J.S., 1965. *Notes on Possible Physical Causes of Long-term Weather Anomalies.* WMO Technical Note 66, World Meteorological Organisation, Geneva, pp. 227–248.

11. Hecht, A.D., 1986: Certificate of Achievement. Namias Symposium, 1986. J.O. Roads, Editor, Scripps Institution of Oceanography Reference Series 86–17. 202 pp.

12. Ratcliffe, R.A.S. and Murray, R., 1970. New lag associations between North Atlantic sea temperatures and European pressure applied to long-range weather forecasting. *Quarterly Journal of the Royal Meteorological Society*, **96**, 226–246.

13. Palmer, T.N., 1986. Gulf Stream variability and European climate. *Meteorological Magazine*, **115**, 291–297; Palmer, T.N. and Sun, Z., 1985. A modelling and observational study of the relationship between sea surface temperature in the north-west Atlantic and the atmospheric general circulation. *Quarterly Journal of the Royal Meteorological Society*, **111**, 947–975.

14. Houghton, D.D., Kutzbach, J.E., and McClintock, M., 1974. Response of a general circulation model to sea surface temperature perturbations. *Journal of Atmospheric Science*, **131**, 857–868; Kutzbach, J.E., Chervin, R.M., and Houghton, D.D., 1977. Response of the NCAR general circulation model to prescribed changes in ocean surface temperature. Part 1: mid-latitude changes. *Journal of Atmospheric Science*, **34**, 1200–1213.

15. Anon., 2008. Forest Fire? Blame it on the ocean. *New Scientist*, **198**, No. 2661, 21 June, p. 22; Liu, Y., 2006. North Pacific warming and intense north-western US wildfires. *Geophysical Research Letters*, **33**, 1–5.

16. Feng, S., Oglesby, R.J., Rowe, C.M., Loope, D.B., and Hu, Q., 2008. Atlantic and Pacific SST influences on medieval drought in North America simulated by the Community Atmospheric Model. *Journal of Geophysical research*, **113**, doi:10.1029/2007JD009347; Anon., 2008. When oceans conspire to dessicate North America. *New Scientist*, 14 June.

17. The NAO has been described in: Morton, O., 1998. The Storm in the Machine. *New Scientist*, No. 2119, 31 January.

18. Dickson, R.R., Lazier, J., Meincke, J., Rhines, P., and Swift, J., 1996. Long-term co-ordinated changes in the convective activity of the North Atlantic. *Progress in Oceanography*, **38**, 241–295.

19. Curry R.G. and McCartney, M.S. 2001. Ocean gyre circulation changes associated with the North Atlantic Oscillation. *Journal of Physical Oceanography*, **31**, 3374–3400.

20.   Keenlyside, N.S., Latif, M., Jungclaus, J., Kornblueh, L., and Roeckner, E., 2008. Advancing decadal-scale climate prediction in the North Atlantic sector. *Nature*, **453**, 84–88.

21.   Rodwell, M.J., Rowell, D.P., and Folland, C.K., 1999. Oceanic forcing of the winter-time North Atlantic Oscillation and European climate. *Nature*, **398**, 320–323.

22.   Hoerling, M.P., Hurrell, J.W. and Xu, T. 2001. Tropical origins for North Atlantic climate change. *Science*, **292**, 90–92.

23.   Taylor, A.H., 2005. A model of variations in the North Atlantic Oscillation. *Geophys. Res. Letts.*, **32**, L24713.

24.   A weakness of the simple model is that it is very difficult to know what numbers to use in describing the processes within each resonator. In the case of Fig. 7.2, the numbers were chosen to give the best possible fit to the observed NAO values, a procedure which demonstrates whether the model can work at all. This procedure leaves open the question of whether an equally good fit could be obtained for any set of entirely random NAO data. The question can only be decided by statistical analysis. Such calculations reveal that a fit to the observations as good as this will occur at most once in every hundred sets of random data tried.

25. Taylor, A.H., 1996. North–south shifts of the Gulf Stream: ocean-atmosphere interactions in the North Atlantic. *Intl. J. Climatol*, **16**, 559–583.

Chapter 8

1.   Reuters, 2007. Octopus finds ancient Korean porcelain trove. Tues. 24 July.

2.   Scarth, A., 1999. *Vulcan's fury: man against the volcano*. Yale University Press.

3.   Head, J. and Bloch, J., 2009. Giant snake fossils point to steamy ancient tropical climate. *Nature*, **457**, 634; Head, J.J., Bloch, J.I., Hastings, A.K., Bourque, J.R., Cadena, E.A., Herrera, F.A., Polly, P.D., and Jaramillo, C.A., 2009. *Nature*, **457**, 715–717.

4.   Winchester, S., 2002. *The Map that Changed the World*, Penguin Books, 338pp.; Bryson, B., 2004. *A Short History of Nearly Everything*, Black Swan, 687pp.

5.   Anon., 2006. The nitrogen the Vikings left behind. *New Scientist*, **191**, 9 September, p. 19.

6.   Geddes, L., 2010. Hidden graves spotted by airborne imaging. *New Scientist*, **206**, 10 April, 18–19.

7.   Mott, M., 2003. Can animals sense earthquakes. *National Geographic News*, 11 November.

8.   Grant, R.A. and Halliday, T., 2010. Predicting the unpredictable: evidence of pre-seismic anticipatory behaviour in the common toad. *Journal of Zoology*, **281**, 263–271.

9.   Krajick, K., 2001. Arctic life on thin ice. *Science*, **291**, 424–425; DeWitt, L., 2006. Flight risk: a bird's eye view of climate change. *Audobon Naturalist Society*, **32**, No. 3.

10.   Both, C., Bouwhuis, S., Lessells, C.M., and Visser, M.E., 2006. Climate change and population declines in a long-distance migratory bird. *Nature*, **441**, 81–83.

11.   Shurin, J.B., 2006. A pitcher of things to come. *Nature*, **443**, 399–400.

12.   Gotelli, N. and Ellison, A.M., 2006. *PloS Biol.*, **4**, e324.

13.   Srivastava, D.S. *Oecologia*, **149**, 493–504.

14.   Moore, P.D., 2006. Green and pleasant trials. *Nature*, **440**, 613–614; Terborgh, J. et al., 2001. *Science*, **294**, 1923–1926; Terborgh, J., Feeley, K., Silman, M., Nunez, P., and Balukjian, B. *Journal of Ecology*, **94**, 253–263.

15.   Stolzenburg, W., 2008. *Where the Wild Things Were: Life, Death and Ecological Wreckage of Vanishing Predators*. Bloomsbury, 240pp.

16.   Scott Mills, L., Soule, M.E., and Doak, D.F., 1993. The keystone species concept in ecology and conservation. *Bioscience*, **43** (4), 219; Paine, R.T., 1995. A conversation on refining the concept of keystone species. *Conservation Biology*, **9** (4), 962–964.

17.   Wilson, E., 1992. *The Diversity of Life*, Harvard University Press.

18.   Pimm, S., 2008. Missing links in food-chain story. *Nature*, **454**, 275–276.

19.   Pearce, F., 2006. The fog catcher's forest. *New Scientist*, **191**, 6 August, 37–39.

20.   Pearce, F., 2006. The miracle of the stones. *New Scientist*, **191**, 9 September, 50–51.

21.   Pearce, F., 2004. The accidental rainforest. *New Scientist*, 18 Sepember, 44–45; Wilkinson, D., 2004. The parable of Green Mountain. *Journal of Biogeography*, **31**, p. 1.

22.   Kaiser, J., 2001. How rain pulses drive biome growth. *Science*, **291**, 413–414; Knapp, A.K. and Smith, M.D., 2001. Variation among biomes in temporal dynamics of above-ground primary production. *Science*, **191**, 481–484.

23.   Helvarg, D. 1999. Elegant scavengers: giant petrels are a bellwether species for the threatened Antarctic Peninsula. *E Magazine*, November/December; Walsh, J.E., 1999. The Arctic as a bellwether. *Nature*, **352**, 19–20.

24.   The history of Croll and Milankovitch's work is described in: Imbrie, J. and Imbrie, K. Palmer., 1979. *Ice-Ages: Solving the Mystery*. Harvard University Press; Bryson, B., 2004. *A Short History of Nearly Everything*. Black Swan.

25.   Kerr, R.A., 2000. Does the climate clock get a noisy boost? *Science*, **290**, 697–698.

26.   Shackleton, N.J., 2000. The 100,000-Year Ice-Age Cycle Identified and Found to Lag Temperature, Carbon Dioxide, and Orbital Eccentricity. *Science*, **289**, 1897–1902;

Kerr, R.A., 2000. Ice, Mud Point to CO$_2$ Role in Glacial Cycle. *Science*, **289**, 1868. Other possible feedback processes are described in: Battersby, S., 2010. The great meltdown. *New Scientist*, **206**, 22 May, 32–36.

27. Sigman, D.M. and Boyle, E.A., 2000. Glacial/interglacial variations in atmospheric carbon dioxide. *Nature*, **407**, 859–860.

28. Cane, M.A., Eshel, G., and Buckland, R.W., 1994. Forecasting Zimbabwean maize yield using eastern equatorial Pacific sea surface temperature. *Nature*, **370**, 204–205.

29. Anon., 2006. Climate killing Africa's crops. *New Scientist*, 26 February, p. 7.

30. Herren, H. and Stenseth, N., 2006. *Proceedings of the National Academy of Sciences*, **103**, 3049.

31. Rosenzweig, C., 1994. Maize suffers a sea-change. *Nature*, **370**, 175–176.

32. Taylor, A.H., Allen, J.I., and Clark, P.A. 2002. Extraction of a weak climatic signal by an ecosystem. *Nature*, **416**, 629–632.

33. Hsieh, C., Reiss, C.S., Hunter, J.R., Beddington, J.R., May, R.M., and Sugihara, G., 2007. Fishing elevates variability in the abundance of exploited species. *Nature*, **443**, 859–862.

34. Sitch, S., Cox, P.M., Collins, W.J., and Huntington, C., 2007. Indirect radiative forcing of climate change through ozone effects on the land–carbon sink. *Nature*, **448**, 791–794. The harmful effects of UV on phytoplankton have also been discussed in: Joint, I.R. and Jordan, M.B., 2008. Effect of short-term exposure to UVA and UVB on potential phytoplankton production in UK coastal waters. *Journal of Plankton Research*, **30**, 199–210.

35. Woodman, R. and Wilson, J., 2004. *The Lighthouses of Trinity House*, Thomas Reed Publications, 256pp.; Medland, J., 2007. The Island's oldest lighthouse. *West Wight Beacon*, July, No. 160.

Chapter 9

1. Uglow, J., 2002. *The Lunar Men: The Friends Who Made the Future*. Faber and Faber, 588pp.

2. Schofield, R.E., *The Lunar Society at Birmingham: A social history of provincial science and industry in 18th-century England*, Oxford: Clarendon Press, 1963.

3. Pearce, F., 2006. The ice age that never was. *New Scientist*, 16 December, 46–47.

4. Christianson, G.E., 1999. *Greenhouse: The 200-year story of global warming*. Constable, London, 305pp.

5. Burroughs, W.J., 2001. *Climate Change: A Multidisciplinary Approach*. Cambridge University Press, 298p; Wood, R.W., 1909. A note on the theory of the green house. *Philosophical Magazine*, **17**, 319–320.

6. Eve, A.S. and Creasey, C.H., 1945. *The Life and Work of John Tyndall*, Macmillan, London, 430p; Pain, S., 2009. A discovery in need of a controversy. *New Scientist*, 16 May, 46–47; Hulme, M., 2009. On the origin of 'the greenhouse effect': John Tyndall's 1859 interrogation of nature. *Weather*, **64**, 121–123.

7. Pearce, F., 2003. Land of the midnight sums. *New Scientist*, 25 January, 50–51.

8. Arrhenius, S., 1896. On the influence of carbonic acid in the air upon the temperature of the ground. *The London, Edinburgh and Dublin Philosophical Magazine and Journal of Science*, 5th series (April), 237–276.

9. Mason, N. and Hughes, P., 1999. *Introduction to Environmental Physics: Planet Earth, Life and Climate*. Taylor and Francis, London and New York, 463p.

10. Charlson, R.J., 2007. A lone voice in the greenhouse. *Nature*, **448**, 254; Fleming, J.R., 2007. *The Callendar Effect: The Life and Work of Guy Stewart Callendar (1898–1964), the Scientist who Established the Carbon Dioxide Theory of Climate Change*, American Meteorological Society, 176pp.

11. Sawyer, J.S., 1972. Man-made carbon dioxide and the 'greenhouse' effect. *Nature*, **239**, 23–26.

12. Nicholls, N., 2007. Climate: Sawyer predicted rate of warming in 1972. *Nature*, **448**, 992.

13. Heimann, M., 2005. Charles David Keeling 1928–2005: pioneer in the modern science of climate change. *Nature*, **437**, 331.

14. IPCC., 1995. *Climate Change 1995: The Science of Climate Change*. Houghton, J.T., Meira Filho, L.G., Callendar, B.A., Harris, N., Fattenberg, A., and Maskell, K. (eds). Cambridge University Press, Cambridge, UK.

15. Anon., 2008. Climate calamity. *New Scientist*, 4 October, 7.

16. Hefferman, O., 2010. The climate machine. *Nature*, **463**, 1014–1016.

17. The theory is presented more fully in: Svensmark, H. and Calder, N., 2007. *The Chilling Stars, A Cosmic View of Climate Change*, Icon Books, 268pp.

18. Brahic, C., Chandler, D.L., Le Page, M., McKenna, P., and Pearce, F., 2007. Climate myths. *New Scientist*, **194**, 34–42. It has recently been suggested that winter temperatures in northern Europe are affected by solar activity, but the causes of such a local effect remain unclear (Clark, S., 2010. Quiet sun puts Europe on ice. *New Scientist*, **206**, 17 April, 6–7; Lockwood, M. et al., to appear in *Environmental Research Letters*).

19. Pearce, F., 2006. Grudge match. *New Scientist*, 18 March, 40–43.

20. Mann, M.E., Bradley, R.S., and Hughes, M.K., 1998. Northern hemisphere temperatures during the past millennium: inferences, uncertainties and limitations. *Geophysics research Letters*, **26**, 759–762.

21. Egan, T., 2002. Alaska, no longer so frigid, starts to crack, burn and sag. *The New York Times*, June 16.

22. de la Mare, W.K., 1997. Abrupt mid-20th century decline in Antarctic sea-ice extent from whaling records. *Nature*, **389**, 57–60.

23. Pain, S., 2009. A discovery in need of a controversy. *New Scientist*, 16 May, 46–47.

24. Rignot, E. and Kanagaratnam, P., 2006. Changes in the velocity structure of the Greenland ice sheet. *Science*, **311**, 986–990.

25. Flannery, T., 2006. *The Weather Makers: The History and Future Impact of Climate Change*, Allen Lane, 341pp.

26. BBC News 21 April 2005. Antarctic glaciers show retreat; BBC News 24 September 2004, West Antarctic glaciers speed up.

27. Levitus, S., Antonov, J.I., Boyer, T.P., and Stephens, C., 2000. Warming of the World Ocean. *Science*, **287**, 2225–2229.

28. Curry, R.G., Dickson, R., and Yashayaev, I., 2003. A change in the freshwater balance of the North Atlantic over the past four decades. *Nature*, **426**, 826–829.

29. Bryden, H.L., Longworth, H.R., and Cunningham, S.A. Slowing of the Atlantic meridional overturning circulation at 25°N., 2005. *Nature*, **438**, 655–657.

30. Cunningham, S.A., et al. 2007. Temporal variability of the Atlantic Meridional Overturning. *Science*, **317**, 935–937; Schiermeier, Q. 2007, Ocean circulation noisy, not stalling. *Nature*, **448**, 844–845.

31. Pain, S., 2006. Carteret's south sea trouble. *New Scientist*, 11 February, 52–53; *Carteret's Voyage Round the World 1766–1769*, his own account of his journey was finally published in 1965 by the Hakluyt Society.

32. Zukerman, W., 2010. Pacific islands defy sealevel rise. *New Scientist*, **206**, 5 June, 10; Kench, P. and Webb, A. *Global and Planetary Change*, doi:10.1016/j.gloplacha.2007.11.001.

33. Anon., 2010. Hotter heatwaves. *Nature*, **466**, 669; Clark, R et al., 2010. *Geophysical Research Letters*, (doi:10.1029/GL043898).

Chapter 10

1. Quoted by Ring, J., 2008. Lunatic cactus. *New Scientist*, 8 March, 81.

2. Wood, R.W., 1904. The N-Rays. *Nature*, **70**, 884–885.

3. Taylor, A.H.. 2002. North Atlantic climatic signals and the plankton of the European Continental Shelf, *Large Marine Ecosystems of the North Atlantic: Changing States and Sustainability* ed. by K. Sherman and H.R. Skjoldal., 3–26.

4. Taylor A.H., Colebrook J.M., Stephens J.A., and Baker N.G.. 1992. Latitudinal displacements of the Gulf Stream and the abundance of plankton in the north-east Atlantic. *Journal of the Marine Biological Association*, **72**, 919–921; Hays, G.C., Carr, M.C., and Taylor, A.H., 1993. The relationship between Gulf Stream position and copepod abundance derived from the Continuous Plankton Recorder Survey: separating biological signal from sampling noise. *J. Plankt. Res.*, **15**, 1359–1373.

5. Taylor, A.H., Allen, J.I., and Clark, P.A. 2002. Extraction of a weak climatic signal by an ecosystem. *Nature*, **416**, 629–632.

6. Vince, G., 2009. Surviving in a warmer world. *New Scientist*, 28 February, 28–33.

7. Pearce, F., 2007. But here's what they didn't tell us. *New Scientist*, 10 February, 7–9; Ananthaswamy, A., 2009. Going, going... *New Scientist*, 4 July, 28–33.

8. Walker, G., 2006. The tipping point of the iceberg. *Nature*, **441**, 802–805.

9. Rahmstorf, S., Crucifix, M., Ganopolski, A., Goosse, H., Kamenkovich, I.V., Knutti, R, Lohmann, G., Marsh, R., Myzak, L.A., Wang, Z., and Weaver, A.J., 2005. Thermohaline hysteresis: a model intercomparison. *Geophysical. Research Letters*, **32**, L23605.

10. Trenberth, K.E. et al., 2002. Evolution of El Niño – Southern Oscillation and global surface temperatures. *Journal of Geophysical Research*, **107** (D8), 4065.

11. Collins, M. and The CMIP Modelling Groups., 2005. El-Niño- or La-Niña-like climate change. *Climate Dynamics*, **24** (1), 89–104.

12. Merryfield, W.J., 2006. Changes to ENSO under CO_2 doubling in a multimodel ensemble. *Journal of Climate*, **19**, 4009–4027. There are indications that the warm water pool in El Niño conditions is becoming displaced further west: Yeh, S.-W., Kug, J.-S., Dewitte, B., Kwon, M.-H., Kirtman, B.P., and Jin, F.-F., 2009. El Niño in a changing climate. *Nature*, **461**, 511–514.

13. http://www.sciencedaily.com/releases/2009/01/090113101200.htm.

14. Taylor, A.H., 2005. A model of variations in the North Atlantic Oscillation. *Geophys. Res. Letts.*, **32**, L24713.

15. Thuiller, W., 2007. Biodiversity: climate change and the ecologist. *Nature*, **448**, August.

16. Sala, O.E. Chapin, F.S. III, Armsest, J.J., Berlow, E., Bloomfield, J., Dirzo, R., Huber-Sanfeld, E., Huenneke, L.F., Jackson, R.B., Kinzig, A., Leemans, R., Lodge, D.M., Mooney, H.A., Oesterheld, M., LeRoy Poff, N., Sykes, M.T., Walker, B.H., Walker, M.,

and Wall, D.H., 2000. Global Biodiversity Scenarios for the year, 2100. *Science*, **287**, 1770–1774.

17.   Feng, Xiahong., 1999. Trends in intrinsic water use of natural trees for the past 100–200 years: a response to atmospheric CO_2 concentration. *Geochimica et Cosmochimica Acta*, **63**, 1891–1903.

18.   Doney, S.C., 2007. Plankton in a warmer world. *Nature*, **444**, 695–696; Behrenfeld, M.J., O'Malley, R.T., Siegel, D.A., McClain, C.R., Sarmiento, J.L., Feldman, G.C., Milligan, A.J., Falkowski, P.G., Letelier, R.M., and Boss, E.S., 2007. Climate-driven trends in contemporary ocean productivity. *Nature*, **444**, 752–755.

19.   Siegel, D.A. and Franz, B.A., 2010. Century of phytoplankton change. *Nature*, **466**, 569–571; Boyce, D.G., Lewis, M.R., and Worm, B., 2010. Global phytoplankton decline over the past century. *Nature*, **466**, 591–596.

20.   Danis, P.-A., von Grafenstein, U., and Masson-Delmotte, V., 2003. Sensitivity of deep lake temperature to past and future climatic changes: a modeling study for Lac d'Annecy, and Ammersee, Germany. *Journal of Geophysical Research*, **108(D19)**, 4609.

21.   Jones, S., 2000. *Almost Like a Whale*, Transworld Publishers, 499pp.; www.robert-marsham.co.uk.

22.   Fitter, A.H. and Fitter, R.S.R., 2002. Rapid changes in flowering plant time in British plants. *Science*, **296**, 1689–1691.

23.   Brown, J.L., Li, S.-H. and Bhagabati, N., 1999. Long-term trend towards earlier breeding in an American bird: A response to global warming. *Proceedings of the National Academy of Sciences*, **96**, May 11, 5565.

24.   Rosenzweig, C., Karoly, D., Vicarelli, M., Neofotis, P., Wu, Q., Casassa, G., Menzel, A., Root, T.L., Estrella, N., Seguin, B., Tryjanowski, P., Lui, C., Rawlins, S., and Imeson, A., 2008. Attributing physical and biological impacts to anthropogenic climate change. *Nature*, **453**, 353–357; Pearce, F., 2008. Life feels the effects of changing climate. *New Scientist*, 17 May, 10.

25.   Ananthaswamy, A., 2003. Rising clouds leave forests high and dry. *New Scientist*, 22 March, 18.

26.   Pounds, J.A., Fogden, M.P.L., and Campbell, J.H., 1999. Biological response to climate change on a tropical mountain. *Nature*, **398**, 611–615.

27.   Early warning signals for tipping points have been discussed by: Scheffer, M., Bascompte, J., Brock, W.A., Brovkin, V., Carpenter, S.R., Dakos, V., Held, H., van Nes, E.H., Rietkerk, M., and Sugihara, G., 2009. Early-warning signals for critical transitions. *Nature*, **461**, 53–59. The desert forecasting schemes are in: Sole, R., 2007. Scaling laws in the drier. *Nature*, **449**, 151–153; Kefi, S., Rietkerk, M., Alados, C.L., Pueyo, Y., Papanastasis, V.P., ElAich, A., and de Ruiter, P.C. *Nature*, **449**, 213–217; Scanlon, T.M.,

Caylor, K.K., Levin, S.A., and Rodriguez-Iturbe, I., 2007. Positive feedbacks promote power-law clustering of Kalahari vegetation. *Nature*, **449**, 209–212.

28.   Lee, E., Chase, T.N., and Rajagopalan, B., 2008. Highly improved predictive skill in the forecasting of the East Asian summer monsoon. *Water Research*, 44, W10422, doi:10.1029/2007/WR006514.

29.   Joppa, L.N., 2009. An ecologist calls for a citizen-science 'Wiki'. *Nature*, **459**, 619.

30.   Schmeller, D., Henry, P.Y., Julliard, D.R., Gruber, D.B., Clobert, H.J., Dziock, F., Lengyel, B.D.S., Kull, J.T., Tali, K.K., Bauch, K.B., Settele, G.J., Van Sway, L.C., Kobler, M.A., Babij, N.V., and Papastergiadou, K.H.E., 2009. Advantages of volunteer-based biodiversity monitoring in Europe. *Conservation Biology*, **23**, 307–316.

31.   Pilkey, O.H. and Pilkey-Jarvis, L., 2007. *Useless Arithmetic: Why Environmental Scientist Can't Predict the Future*. Columbia University Press; Pearce, F., 2007. No way to save the planet. *New Scientist*, 17 March, 53.

32.   Anon., 2008. There's no 'p' in climate change. *New Scientist*, 19 July, 17; Brikowski, T.H., Lotan, Y., and Pearle, M.S., 2008. Climate-related increase in the prevalence of urolithiasis in the United States. *Proceedings of the National Academy of Sciences*, **105**, 9836–9841.

33.   Harte, J., 2008. An ecologist notes that important details are missing from climate-change models. *Nature*, **454**, 1033; Björk, R. and Molau, U., 2007. *Arctic Antarctic Alpine Research*, **39**, 34–43.

GLOSSARY

Azores high—A large subtropical semi-permanent centre of high atmospheric pressure found near the Azores in the Atlantic Ocean. It forms one pole of the North Atlantic oscillation with the Iceland low.

Bibury roadside verge data set—A series of counts of the number and species of wild plants present on the verge of the old Roman road of Akeman Street near the UK Cotswold village of Bibury. The observations have been made annually since the 1960s by a group led by Arthur Willis.

Blocking—A phenomenon, most often associated with stationary high pressure systems in the mid latitudes of the northern hemisphere, which produces periods of abnormal weather.

Bottom water—The lowermost water mass of a body of water, with well-defined physical, chemical, and ecological characteristics.

BRS model (of the latitude of the Gulf Stream's north wall)—Based on a model developed by Behringer, Regier, and Stommel, this model predicts the latitude of the Gulf Stream by representing the North Atlantic as a single north–south line of boxes. The winds employed are derived from the NAO index, using the observed tendency for the wind system to be displaced north or south as the NAO rises or falls in intensity. Movement of water and redistribution of heat have the effect of integrating the winds and introducing a delay.

Calanus—A small copepod crustacean less than half a centimetre long that is extremely abundant at certain times of the year in the North Atlantic and the North Sea. It is one of the principal foods of fish such as herring and mackerel, and also of the Greenland whale.

California Cooperative Oceanic Fisheries Investigations (CalCOFI) programme—A survey of larval fish in the waters off the shores of California begun in the 1940s when the sardine populations collapsed.

Climatic seesaw—Large-scale swaying in the global weather system, first described by Sir Gilbert Walker, in which the surface atmospheric pressure in one region goes up or down at the same time that, in another region, the pressure does the opposite. Two of the most important are the El Niño Southern Oscillation and the North Atlantic Oscillation.

Coccolithophorida—A group of phytoplankton that construct a calcareous shell of round platelets, which may accumulate on the ocean floor as sediment, ultimately contributing to limestone such as chalk.

Continuous plankton recorder (CPR)—A device, invented by Sir Alister Hardy, for sampling plankton in the surface waters of the sea; it is towed behind merchant and other non-research ships.

Continuous plankton recorder survey—A monthly monitoring of the surface plankton in the North East Atlantic and the North Sea carried out using ships of opportunity since 1948.

Copepod—A large subclass of crustaceans all less than 1 cm long—and most of them much smaller. Many are free-swimming members of the plankton. They are one of the most important links in the marine food chain, grazing on phytoplankton and thereby making plant foods available to the many young fishes and other creatures that feed on them.

Coriolis effect—An inertial force described by the 19th century French engineer–mathematician, Gustave-Gaspard Coriolis, in 1835. Coriolis showed that, if the ordinary Newtonian laws of motion of bodies are to be used in a rotating frame of reference, an inertial force—acting to the right of the direction of body motion for anticlockwise rotation of the reference frame or to the left for clockwise rotation—must be included in the equations of motion. The effect of the Coriolis force is an apparent deflection of the path of an object that moves within a rotating coordinate system. The object does not actually deviate from its path, but it appears to do so because of the motion of the coordinate system.

Correlation coefficient—The Pearson correlation coefficient (there are others) is a statistical measure of how similar two data sets are. If two variables go up and down together, the coefficient has a value of 1, if they are unrelated the value is zero, and if they change in exact opposition the value is –1.

Daphnia—Small, planktonic crustaceans, between 0.2 and 5 mm in length. *Daphnia* are members of the order Cladocera, and are one of the several small aquatic crustaceans commonly called water fleas because of their jerky swimming style with limbs moving together (although neither fleas nor parasites). They live in various aquatic environments ranging from acidic swamps to freshwater lakes, ponds, streams, and rivers.

Deterministic chaos—This happens when systems that are deterministic—meaning that their future behaviour is fully determined by their initial conditions with no random elements involved—are highly sensitive to their initial conditions, thereby rendering long-term predictions impossible.

El Niño Southern Oscillation (ENSO)—A quasi-periodic climate pattern that occurs across the tropical Pacific Ocean on average every three to seven years. It is characterised by variations in the temperature of the surface of the tropical eastern Pacific Ocean—warming or cooling known as *El Niño* and *La Niña* respectively—

and air surface pressure in the tropical western Pacific—the *Southern Oscillation*. The two variations are coupled: the warm oceanic phase, El Niño, accompanies high air surface pressure in the west Pacific, while the cold phase, La Niña, accompanies low air surface pressure in the west Pacific. ENSO causes extreme weather such as floods, droughts, and other weather disturbances in many regions of the world. Developing countries, dependent upon agriculture and fishing, particularly those bordering the Pacific Ocean, are the most affected. In popular usage, the El Niño Southern Oscillation is often called just 'El Niño'. El Niño is Spanish for 'the boy' and refers to the Christ child, because periodic warming in the Pacific near South America is usually noticed around Christmas.

Epilimnion—The topmost layer in a thermally stratified lake. It is warmer and typically has a higher dissolved oxygen concentration than deeper waters. Being exposed at the surface, it typically becomes turbulently mixed as a result of surface wind mixing. It is also free to exchange dissolved gases such as oxygen and carbon dioxide with the atmosphere. Because this layer receives the most sunlight, it contains the most phytoplankton. As they grow and reproduce, they absorb nutrients from the water. When they die, they sink into the hypolimnion resulting in the epilimnion becoming depleted of nutrients.

Euphotic zone—The depth of water in a lake or ocean that is exposed to sufficient sunlight for photosynthesis to occur. The depth of the euphotic zone can be affected greatly by the turbidity of the water. It extends from the atmosphere–water interface downwards to a depth where light intensity falls to 1 per cent of that at the surface, so its thickness depends on the extent of light attenuation in the water column. Typical euphotic depths vary from only a few centimetres in highly turbid or very productive lakes, to around 200 metres in the open ocean.

Gaia hypothesis—This holds that biological processes interact with the Earth's climate in such a way as to produce a life-sustaining environment.

Global oceanic conveyer belt—A unifying oceanic circulation that connects the ocean's surface and the thermohaline (deep water) circulation regimes, transporting heat and salt on a planetary scale.

Gulf Stream connection—The tendency for the abundances of some biological populations in the North Sea, north-east Atlantic and the British Isles to track the north–south movements of the Gulf Stream close to the US coast (the GSNW index).

Gulf Stream north wall (GSNW) index—A measure of the latitude of the Gulf Stream between 65°W and 79°W, derived from observations of the position of the north wall of the current using principal components analysis.

Hypolimnion—The dense, bottom layer of water in a thermally stratified lake, lying below the thermocline. Typically the hypolimnion is the coldest layer of a lake in summer. Being at depth, it is isolated from surface wind mixing during summer, and usually receives insufficient irradiance (light) for photosynthesis to occur. In deep, temperate lakes, the bottom-most waters of the hypolimnion are typically close to 4°C throughout the year. The hypolimnion may be much warmer in lakes at warmer latitudes.

Iceland low—A semi-permanent centre of low atmospheric pressure found between Iceland and southern Greenland and extending in the northern hemisphere winter into the Barents Sea. In summer it weakens and shrinks. It is associated with frequent cyclone activity and forms one pole of the North Atlantic Oscillation, the other being the Azores high.

Little Ice Age—A period of cooling that extended from the 16th to the 19th centuries, marked by more frequent cold episodes in Europe, North America and Asia, during which mountain glaciers, especially in the Alps, Norway, Iceland and Alaska expanded substantially. It is generally agreed that there were three temperature minima, beginning about 1650, about 1770, and 1850, each separated by intervals of slight warming.

May's paradox—Randomly generated food webs decrease in stability as they increase in complexity, whereas real food webs are complex.

Mixed layer—An oceanic or limnological layer in which active turbulence has homogenised some range of depths. The surface mixed layer is a layer where this turbulence is generated by winds, cooling, or processes such as evaporation or sea ice formation, which result in an increase in salinity.

North Atlantic deep water (NADW)—A water mass formed in the North Atlantic Ocean, largely in the Labrador Sea and by the sinking of highly saline, dense overflow water from the Greenland Sea. The water mass can be traced around the southern end of Greenland and then, at a depth of 2000–4000 metres, down the coast of Canada and the USA, where it turns east.

North Atlantic Oscillation (NAO)—The dominant pattern of variation of the climate of the North Atlantic Ocean. Its strength is measured by the NAO index which is the difference in standardised atmospheric pressure in winter between the Azores high (the Azores, Lisbon, or Gibraltar) and the Iceland low (Iceland). In winter this index tends to switch between a strong westerly flow with pressure low in the north and high in the south, and a weaker opposite pattern. The former tends to produce warmer temperatures over much of the western North Atlantic region, the latter the reverse.

Pacific North American pattern—A large-scale weather pattern with two modes, denoted positive and negative, and which relates the atmospheric circulation pattern over the North Pacific Ocean with the one over the North American continent.

Paradox of the plankton—The situation where a limited range of resources (light, nutrients) supports a much wider range of planktonic organisms. The paradox stems from a result of the competitive exclusion principle (sometimes referred to as Gause's law), which suggests that when two species compete for the same resource, ultimately only one will persist and the other will be driven to extinction. The high diversity of phytoplankton stands in contrast to the limited range of resources for which they compete with one another (e.g. light, nitrate, phosphate, silicate, iron).

Phytoplankton—The microscopic plants in the plankton that are the basis of the food webs in oceans or lakes.

Plankton—Derived from a Greek word that means 'wanderers', the plankton refers to all the minute plants and animals that live at or near the surface of oceans or lakes. Some are wholly passive and drift to and fro with the currents while others are able to swim about actively. All, however, are subject to the movements of the surface waters. Many sea creatures, such as starfish, crabs, and fishes, are temporary members of the plankton during infancy.

Primary productivity—the rate of production of phytoplankton.

Principal components analysis—A statistical technique for finding the common pattern among a series of observations.

Rossby wave—A large, slow-moving, planetary-scale wave generated in the upper atmosphere by ocean–land temperature contrasts and topographic forcing (winds flowing over mountains), and affected by the Coriolis effect due to the earth's rotation. They are characterised by a long wavelength (about 6000 km), large amplitude (about 3000 km) and slow movement, which is commonly westward relative to the Earth. Rossby waves have also been observed in the ocean. Here they have a wavelength of a few hundred kilometres and nearly always move westward. Rossby waves are also known as planetary waves.

Seesaw (climatic)—See *climatic seesaw*.

Spring bloom—The rapid growth of phytoplankton in oceans or lakes with the return of plentiful solar radiation and warmth in the spring.

Teleconnection—Near simultaneous climatic link between two or more regions, usually deriving from climatic changes in a base region.

Thermocline—A thin but distinct layer in a large body of fluid (e.g. water, such as an ocean or lake, or air, such as an atmosphere), in which temperature changes more rapidly with depth than it does in the layers above or below. In the ocean, the thermocline may be thought of as an invisible blanket which separates the upper mixed layer from the calm deep water below. Depending largely on season, latitude and turbulent mixing by wind, thermoclines may be a semi-permanent feature of the body of water in which they occur, or they may form temporarily in response to phenomena such as the heating of surface water during the day. Factors that affect the depth and thickness of a thermocline include seasonal weather variations, latitude, and local environmental factors.

Thermohaline circulation—The part of the large-scale ocean circulation that is driven by global density gradients created by surface heat and freshwater fluxes. The adjective thermohaline derives from *thermo-* referring to temperature and

-haline referring to salt content—factors which together determine the density of seawater.

Vorticity conservation—Vorticity is a measure of rotation or spin, analogous to angular momentum. The sum of the vorticity due to motion on the earth's surface and that due to the rotation of the earth (the absolute vorticity) has to stay constant.

Water-fleas—See *Daphnia*.

Western boundary currents—Warm, deep, narrow, and fast-flowing currents that form on the west side of ocean basins. Their eastern counterparts are much slower. They carry warm water from the tropics poleward. Examples include the Gulf Stream, the Agulhas Current, and the Kuroshio. This westward intensification of boundary currents is caused by the strengthening of the Coriolis effect with latitude. Because of the shape of the globe, the Coriolis effect is stronger in the latitudes of the westerlies than in the latitudes of the trade winds.

Zooplankton—The small animals in the plankton.

INDEX

Thuiller, Wilfried 229
TIGER Project 71
Tilman, David 95
tilting of thermocline 62, 135
tipping points 142, 225, 226
toads 172, 238
Todd, Charles 128, 129
top predators 174–6
trade winds 28, 32, 45, 112, 113, 116, 132–5,
    149, 177, 274
transitory climatic phenomena 186, 220
Treatise on Limnology 55
Tsonis, Anastasios 142
turbulence 88, 89, 272
Turk, Stella 16
twentieth-century temperature
    rise 204–8
Tyndall, John 193, 194, 197, 198, 212, 213,
    216
    atmospheric water vapour 194
    effect 194

UK Admiralty Research Laboratory 37
UK Meteorological Office 49, 153, 158
ultraviolet light 20, 173, 187
Uncertainty Principle 89
University
    of Alaska 172, 209
    of Amsterdam 95
    of Aukland 216
    of Buenos Aires 230
    of California 242
    of California, Santa Barbara 232
    of Cambridge 182
    of Gothenburg 242
    of Hull 20
    of Maryland 134
    of Massachusetts, Dartmouth 38
    of Minnesota 95
    of Nebraska 154
    of Oslo 59, 119
    of Oxford 10, 92, 110
    of Southampton 214
    of Texas 16, 242
    of Toronto 169
    of Vermont 174
    of Virginia 207, 240
    of Wisconsin 240
    of Wisconsin-Milwaukee 142

    of York 236
Upwelling 131, 135, 136, 233
US Department of Agriculture's Forest
    Service 154
US Geological Survey 171
US National Oceanic and Atmospheric
    Administration 5, 32
US National Snow and Ice Data
    Centre 211, 212
US Navy Oceanographic Office 5, 29
Utrecht University 239

Van Helmont, Jan Baptista 88
vegetation and soil 75, 76
Venice 48
Veronis, George 37
vertical migration of zooplankton 66
Vikings 28, 171
volcanic stones 177
volcanoes 74, 134, 168, 177, 192, 198, 208
volunteer observers 241
von Grafenstein, U. 234
von Humboldt, Alexander 131
vorticity 35–7, 150, 274
    absolute 35, 116
    planetary 35, 36, 116
    relative 35, 36
Vostock Station, Antarctica 182, 202

Waddensea Island sand dunes 97
Wainwright, Alfred 245
Walker, Sir Gilbert T. 122, 126, 129–31
    ENSO 138, 139, 143, 269
    NAO 130, 139, 143
    Walker circulation 130, 132
warmer ocean 133, 154, 238
Warren, M. S. 17
Washington State University 96
water-fleas 67, 270
Webb, Arthur 216, 217
Webster, John 218, 219, 220, 222
westerly winds 32, 35, 46, 62, 113, 115,
    116, 122, 123, 149, 150, 226
    NAO 157, 272
western boundary currents 34, 40,
    274
Whan-suk, Moon 166
wildfires 154, 155
Willis, Arthur 69–71, 76, 247, 269